THE
FUTURE
MAKERS

THE FUTURE MAKERS

IN DIGITAL TECHNOLOGY

EDITED BY
KATH WALTERS

First published in 2018 by Kath Walters
246 Amess Street, Carlton North 3054

Typeset in Minion Pro, Gotham Book and Gotham Bold by Lu Sexton.
© Kath Walters 2018

The moral rights of all the authors and the editor have been asserted.

All rights reserved. Except as permitted under the *Australian Copyright Act 1968* (for example, a fair dealing for the purposes of study, research, criticism or review), no part of this book may be reproduced, stored in a retrieval system, communicated or transmitted in any form or by any means without prior written permission. All inquiries should be made to the publisher at the above address or by email to kath@kathwalters.com.au

Cover design by Lorna Hendry
Page design by Jen Clark
Printed by Ingram Spark
Editor's photo by Beth Jennings

ISBN 978-0-6482928-0-7 (print edition) 978-0-6482928-1-4 (e book)

Disclaimer:
The material in this publication is of the nature of general comment only and does not represent professional advice. It is not intended to provide specific guidance for particular circumstances and it should not be relied upon for any decision to act or not to act on any matter which it covers. Reader should obtain professional advice where appropriate, before making any such decision. To the maximum extent permitted by law, the author and publisher disclaim all responsibility and liability to any person, arising directly or indirectly from any person taking or not taking action based on the information in this book.

For my daughter, Audrey Foley, who is a daily inspiration.

GRATITUDE

First and foremost, my thanks goes to the authors – the Future Makers – who have contributed with their heads, hearts and hands to the creation of this book. It takes guts to put controversial ideas on paper. You have been thorough, diligent, supportive and committed from the start and I am deeply grateful.

My thanks also goes to my daughter, Audrey Foley, who helped in untold ways to bring this book to fruition. Thanks to my designers, Jen Clark and Lorna Hendry, my typesetter, Lu Sexton, and my assistant, Nicole McConville, for her brilliant scheduling. Every one of you has gone above and beyond duty in supporting this project. Also, to my dear friend, Bernadette Foley, who helped me find the focus and title of this book.

I owe much to the founders of the Thought Leader Business School, Matt Church and Peter Cook, in particular for alerting me to the power of self-publishing in the digital age.

Mentors have played a big role in my life, in particular, Amanda Gome and Yamini Naidu, who are also both dear friends. I'm truly lucky to have many more wonderful friends who come along on my life's journey. You know who you are, and I thank you for the many ways you enrich my life.

CONTENTS

CHAPTER ONE: Lead the Future — 12
By Kath Walters
Future makers are leading the future. How do they do it?

CHAPTER TWO: The "Social Cell Network" — 29
By Alan Lloyd
How a new online model of engagement, collaboration and co-creation of shared value will change our world

CHAPTER THREE: Pitch Big — 45
By Sam Lanyon
The sad truth about why most products fail, and what to do about it

CHAPTER FOUR: Towards Purposeful Companies — 59
By Tom Dawkins
How technology shifts the playing field

CHAPTER FIVE: Directing Change — 81
By Susan Oliver
The strategic role of the director in the digital world

CHAPTER SIX: The Spin In — 101
By Joe Ward
Why our innovation models fail us and how to change them

CHAPTER SEVEN: Digital Technology: It's an Action Sport — 117
By Peter Williams
How do you become a future maker?

CHAPTER ONE

LEAD THE FUTURE

FUTURE MAKERS ARE LEADING THE FUTURE.
HOW DO THEY DO IT?

BY KATH WALTERS

Author's mentor and founder of
www.kathwalters.com.au

OVER MY 20 YEARS AS A BUSINESS journalist, I've spoken to hundreds of company leaders and experts – men and women (but mostly men, which I will address later in this chapter) – who I call 'Future Makers'.

What is a Future Maker? Here's what they have in common. Each of them is crystal clear about what they think and believe. They are hell-bent on changing the way we get stuff done in the future. They have extraordinary talent and insight to offer the world in the form of the goods and services they provide. They lead the future.

Typically, they have been in their industry for five to 10 years, or more. They've seen how it all works, and what doesn't work. Future Makers have got to a pinnacle, where they're looking out in front of them at all the possibilities, and they're also looking behind them at all the people they need to connect with to make the future that they can see become a reality.

But all Future Makers face a similar problem: the need to educate their market.

> *Every Future Maker I have ever spoken to is so far ahead of their market that they need to find a way to bridge the gap between their ideas and expertise and their market's knowledge and understanding.*

That means spending vast amounts of time having conversations just to get prospective customers up to speed. And then, after that discussion, they might find that the person they are talking to is not 'their people' after all.

I once interviewed Peter Farrell, the founder of one of Australia's most successful medical devices company, Resmed, which sells a breathing machine for people with sleep apnea. In 1989, when Farrell founded Resmed, most people dismissed sleep apnea as 'snoring' – a bit of a joke. But Farrell knew it made some sufferers' lives a misery. Their 'snoring' was so bad that they fell asleep at work, in the middle of conversations, even behind the wheel. In preparing for the interview, I went back to Resmed's early annual reports and found Farrell's earliest efforts to educate his market about the real risks of sleep apnea. To create the global success story that is Resmed, Farrell spent almost three decades relentlessly telling the same story again and again. Sleep apnea isn't a joke. It's not just snoring. It is a debilitating and sometimes life-threatening condition. Resmed is now the world's foremost authority on managing the condition.

Future Makers need a platform to express themselves. They need to give their market information *before* they can even start a conversation with them. In the past, one of the few ways Future Makers could educate their market was getting into the media, (which is how I had the great good fortune of speaking to so many Future Makers). They had to get into magazines and newspapers, like *Business Review Weekly* and the *Australian Financial Review*, to become recognised as authorities and experts in their field – as Future Makers.

Farrell was a master of taking his message to the media, winning coverage and building Resmed's success. Like all Future Makers, Farrell was willing to invest as much time and money and effort as needed to make the future he envisioned to happen.

Today, Future Makers have a secret weapon that Farrell didn't have: the opportunity to write and publish a book.

IF YOU WANT TO LEAD THE FUTURE, WRITE A BOOK

By writing and self-publishing a book, a Future Maker encapsulates their intellectual property and stakes their place in the market. They make their ideas robust and relevant to the market. They attract opportunities by becoming known as the expert they are and recognised in their market. They open doors and take their important message to a broader audience.

If you want to lead the future, if you are a Future Maker, the single most valuable focus for your time and effort is capturing your ideas in a book. There are a lot of reasons why.

A bridge between you and your market

As a Future Maker, you are ahead of your market. They are way behind you. Your book is the bridge that links back to the client base you need, your ideal client base.

The author Robyn Haydon's first book, *The Shredder Test: A Step by Step Guide to Writing Winning Proposals*, is all about how to write a winning tender document. Tenders are not sexy documents, but for the people in global engineering or accounting firms, for example, hundreds of jobs can depend on the success of their tenders. So, as you can imagine, the people who write bids are concerned about getting it right.

In her corporate career, Haydon became excellent at writing tenders. She then decided to have a baby, and thought to herself, 'What can I do while my baby is little?' She decided she'd write a book about how to write a tender that doesn't end up in the shredder. She wrote and published this book over 10 years ago – long before this more recent trend to self-publish – so she was ahead of her time.

I've interviewed Haydon, and she told me that within a couple of years that book had increased her revenue as a business consultant by $120,000 a year, and has done so consistently in the years that have followed. Her book was a godsend to her market. People who were struggling to write winning tenders bought her book on the internet. It sold well, and everyone was talking about it. Suddenly, she was the expert and the authority on how to write winning tenders.

In this book, the Future Makers are challenging us to think differently about digital technology. They are building a bridge and saying, 'Come. Come with me.' And that's what Haydon did.

New opportunities

Haydon discovered that her book opened doors to new business, such as training people on the techniques that she describes in that book, which is how she can generate the extra revenue each year.

Until Haydon encapsulated her ideas in that book, people didn't realise that she was an expert and was able to train them. She explained why their tenders weren't working, and then she showed them what a good bid looks like, and how to create a winning tender. They were so excited about that; they'd get her in to train them. The book told them everything they needed to know, but Haydon is still being asked to go in and educate people in applying her principles.

There is a valuable lesson here. People sometimes worry about giving away all their ideas in their book. But even when you are generous in a book, you explain the reasoning behind your thoughts, what they are and how to apply them, people will still want your help. They want you to coach them, or mentor or train them. They want to work more closely with authors, or to buy their products or services.

Clarity, conviction and authority.

In the process of writing, Future Makers gain new clarity about what they have to offer. My clients tell me the process of writing a book delivers clarity, confidence and quality to their ideas that surpass anything they have felt before. This translates into conviction in the marketplace. And the energy, excitement and determination you need to complete your manuscript will convert into dollars. And when you publish, you unleash an authority that you deserve.

Return on investment

Critics might say that writing and publishing a book is an expensive way to win sales. It's true that, depending on the amount of help you need, your book might cost as much as $15 or $20 a copy to produce, and that doesn't include your time (although it takes far less time than you probably think).

Here's what I think. Your book has the potential to make you 10 times that investment. If it cost you $12,000 to publish your book and you sell 12 training workshops as a result for $10,000 each, that is what I call a best seller: a $120,000 return. So write a book for a market that can deliver you a 10 times return.

WHY DON'T YOU JUST ...

Not everyone agrees with me about the power and purpose of writing a book.

Critics of my ideas might say that self-publishing is a 'vanity', that if you can't get a publisher, it means your book is not good enough. These critics don't understand the landscape of traditional publishing. Publishers have constraints, budgets, and timetables. They can see the best book in the world and just not be able to add it to their list because their list is full, or the market is not big enough, or it isn't the kind of book they publish.

Others might tell you just to get somebody to write it for you – a ghost-writer – so you don't have to do this work yourself. I am going to quote Robyn Haydon here because she addresses this point so eloquently. She says, 'No one can sell you as you sell yourself.' Future Makers need to write their books, but that doesn't mean doing so without any guidance and help. It's a simple matter of working with people who can help you structure your ideas to create the quality. I'm not the only person who can do that. There's a whole industry of people out there in the market who can help you write, edit and self-publish a book.

Some might argue that you're better to spend your money on logos, or websites, or buying mailing lists of companies to contact. I disagree. The best quality leads come from writing and publishing a book. If you want to get qualified leads – people who want to work with you because of the unusual attributes you bring to the world – the single most valuable thing you can do is to encapsulate your ideas in a book. Of course, you need a website. But the trend in websites today is to keep them simple, like a brochure. You can't communicate the depth of your ideas and their quality or the experience behind them on your website. In fact, nobody wants to read your website. They want a quick overview. Your website is not where you find quality leads.

WHAT IS A BOOK?

My definition of a Future Maker's book is 25,000 to 45,000 words about your area of expertise. It's nine chapters in three parts: why, what and how, and is published in a small quantity of 100 or 200 copies using digital printing. I am not talking about writing e-books in this chapter. An e-book version of your book is essential, but the power and authority I am talking about come from printed books.

The first section of this book is about is 'why' it is important to read your book. You name the problems that your readers are struggling with. You meet them with 'why'. As the author Simon Sinek says in his bestselling book: *Start with Why*.

The next section of your book deals with 'what' you are talking about. In this part, you make distinctions between how most people think about the problems you have named, and what you think about them. What is different about the future that you want to make? Canvas and consider other people's point of view and answer those critics. Challenge the reader by saying, 'Look; I know you've got these problems. I know you've been tackling them your way, but it's not working, so have a look at my way.'

The final part of your book is all about 'how' to do what you say. You must show your readers how to solve their problems because you have taken them on a journey. You have named their issues and challenges, so you must inspire them with a solution. You must leave them with a sense of how to change or how to solve their problems the way you suggest.

HOW TO WRITE A BRILLIANT BUSINESS BOOK IN 90 DAYS

Once you realise that writing a book is the ultimate positioning tool, the ultimate tool to build your authority, you'll become determined to do it. But where do you start? Let me show you.

I'm going to give you all the steps here. In fact, the steps over the page shows my whole book program in a nutshell.

> *Using the ideas below, it shouldn't take you more than about 40 hours to write an entire book. Oh, and by the way, if you are writing just a chapter in a book, like the contributors to this book, you can use the same process.*

I break down my book process into three stages: Focus, Create and Publish. However, you can only use this method if you commit to the Why-What-How structure. I have seen so many would-be authors get caught in the never-ending process of restructuring their ideas this way, and that way. The Why-What-How structure is not unique. In fact, it is tried and true. Even the bestselling author, Stephen Covey's, in *The Seven Habits of Highly Effective People,* follows this structure. Sure, the 'how' section is the longest and is divided into the seven numbers, but there are what and why parts too. (I checked.)

Also, before you start, surrender to one truth: every book, even a business book, is a story. For every idea you have, for every message you want to convey, find a story that sums it up and illustrates it. And then set yourself a challenge: can you say everything you want to say using stories, anecdotes, examples or metaphors. Stories will bring your messages alive.

Stage one: Focus

Audience first

'Focus' is the first stage. Shift your focus from writing – whether you are good at it or not – to your audience. You need to identify the audience, the reader that you want to reach with your ideas. I insist that my clients pick a person that they have worked with, an ideal client they wish there were more of in the world. Their perfect client becomes the reader. We develop a detailed profile of that reader. Write out a description of their background, age, gender, education, career level, sexual preference, family life and so on.

Once you've got that clarity about the reader of your book, the next step is so much easier: which ideas do you need to convey to that particular reader? You get incredible clarity on your ideas here. For example, if you are writing a book for early-career entrepreneurs, the ideas that you would share with them would be different from the ideas that you might share with a chief executive. The audience determines the ideas that you're going to write about.

Now, look at their problems. What are their three most significant challenges? What causes these issues and what is the impact of not solving them. Are they stressed, anxious, frustrated? Write all this down. Get your audience and their problems crystal clear because once you do, you are ready to write your book.

Stage two: Create

Structure your ideas into a book outline

The three main problems your reader faces, their causes and impacts, will become the first three chapters of your book. These problems are the reason 'why' your reader wants to read your book. You write a whole chapter about each problem because these problems are causing your reader so much pain that they are willing to read your book about how to solve them. So, once you have outlined the first three chapters, what comes next?

Three chapters about 'what'. What do you think that is different about each of those problems? For example, if people have a problem with conflict, they have probably been advised how to solve a conflict. What is different about your ideas on conflict? Perhaps you think conflict is creative when others say it is destructive. Your next three chapters address what you think about their problems that are different from other Future Makers. You now have six chapters in your outline.

The last three chapters will tell your readers 'how' they can do what you say. For example, my message in this chapter is to write a book. I've told you that, as a Future Maker, you have a problem. I've told you what you need to do: write a book. Now I am telling you how to write a book.

Flesh out your book outline into your chapters
Now flesh out your chapters. Explain the point of each chapter and why it is essential. Use stories to illustrate the benefit of understanding your point, clarify vital distinctions in what you are talking about, and show how they apply to their circumstances. Do you have a short survey or diagnostic to help them understand your point? Myers Brigg is a diagnostic that helps people understand the problems they have as a result of their particular personality. Your diagnostic might be much more straightforward. I interview my clients to help them create their chapters. I ask them 18 questions, and when they answer those questions, they have created a chapter. Just in answering those questions, the chapter is written.

Add your introduction and conclusion
In your introduction, describe the readers you are writing for and explain why you are qualified to provide an answer to them. In your conclusion, remind them to take action, name the barriers they might encounter and then invite them to contact you.

Stage Three: Publish

Publishing is a process of letting go. We all feel vulnerable when we put our ideas down on paper. Will others criticise us? (Yes!) Will we build our brand or destroy it? (Ouch). Women are especially prone to this kind of anxiety. Yes, a massively sweeping statement but I've got some evidence. I asked dozens of men and women to contribute their ideas to this book, and only one woman accepted the challenge. So that's me throwing down the gauntlet to all you women who are Future Makers.

Back to the process of publishing: it's a matter of passing your book through several hands to polish it. First, ask a friend to read it. Be specific about what you want them to comment on. Am I articulating my ideas with clarity? Is it building my authority? Is it going to be the bridge that I need to bring the right clients into my world?

Then commission an editor to review your work. Is it concise? Are your ideas clear? Then send it to a proofreader to check your spelling and grammar. Now it is ready for design. Recheck it before you print it. Believe me, once you have been through this process, you will be prepared to publish your book. You may have to make some revisions. Maybe many changes. Then it is ready to print.

Publishing reminds me a bit of giving birth (kind of). I dreaded labour all through my pregnancy, but when my daughter was two weeks overdue, I quickly changed my view. I was desperate to get on with it. The whole process of editing, proofing, designing and printing will prepare you to give birth to your book.

No one creates a book alone. When we put our words down on paper or even on the web, we can be left feeling vulnerable. We are sticking a stake in the sand. We're saying, 'Hey, look what I think.' We need to have a process of letting go of that fear, and that is to have people who work with you towards this end.

Two apps that help

Rev

Rev is a recording app with a transcription service attached that, at time of printing, cost $1 a minute, which means an hour of recording costs $60 to transcribe. For most people, an hour of speak equals about 4000 to 5000 words – a chapter's worth in a 45,000-word book.

I recommend recording as a way to smooth the writing process. I wrote this chapter by first recording answers to a series of questions and then getting a transcription because this is the quickest way to get an early draft down on paper. And it's much easier to work with words on paper – provided you use a clear structure – than it is to work with a blank page. The recording for this chapter was done in just over an hour.

Also, just chatting naturally with the interviewer helps overcome a writer's blocks. When we write, particularly something as ambitious as a book, it's tempting to try to make each sentence perfect from the beginning. This is impossible, so we end up getting nowhere. Another advantage of recording is that it keeps our 'writing' conversational. One of the pitfalls for inexperienced writers is becoming extremely formal when writing and using lots of passive sentence constructions and nominalisations (turning verbs into nouns). We don't usually speak like that. This process of the interview, working with a buddy, recording it and transcribing it, helps remove this writing flaw. And not only does the interview process help writers to articulate their thoughts quite naturally, but it helps the reader of their business book to understand what they're reading comfortably. Experts can quickly fall into language that is hard work for their reader.

Grammarly
When you get to the point of reviewing and publishing your work, after the second draft just before you give your manuscript to your first reader, there's a fantastic app called Grammarly. Grammarly is a thorough grammar and spelling checker – stronger than you would find in standard word processing software. It's free, but I recommend the paid version because it has a fantastic plagiarism feature. It's quite easy to plagiarise accidentally, and Grammarly helps prevent that. I am just about to run this chapter through Grammarly before I send it to my designer.

ARE YOU READY TO LEAD THE FUTURE?

Do you have great ideas, products or services that you want to share with the world? Do you want to open the door to new opportunities and increase your impact, credibility and fees? Are you ready to be recognised by the market as the expert you are?

Write a book and everybody will know who you are and what you stand for. If you want to change the world, write a book. You know what the future looks like, and writing a book will help you to lead other people towards that future. The world needs excellent ideas beautifully expressed and shared.

Imagine your day starts with people contacting you via email, by phone, wanting to meet you. They are the people that you love working with and spending time with. They have read your book and have glimpsed the future that you can see so clearly. Just talking with them is exciting for you. They deeply understand you, your ideas and your approach. They're full of respect for you and admire the work that you've done to create your book and express your ideas. They see you as a guide, who is capable of leading the future, a person with the ability and the skills to take them on a vitally important journey towards something super valuable in their world.

For some of your day, you talk to the best and most valuable prospective clients you could imagine meeting. The rest of your day, you spend working with clients who are fully committed to working with you in a creative, dynamic partnership to build this exciting future.

That is your future as a self-published author.

The book that you are about to read is a collection of the Future Makers that I've been talking about. They have all taken the time and the effort to capture their ideas in the chapters of this book, and we have self-published them as a group. These are the people looking ahead from the pinnacle and looking behind to the people that they need to bring along on the journey.

Instead of despairing about the distance between them and the people behind them, they have put their heart and soul and their intellect into creating these chapters to build a bridge and to help to lead a future that is full of the potential they see.

I'm excited for you to read these chapters ahead. Let me introduce you to the Future Makers. You will be challenged and inspired by their ideas.

GET IN TOUCH

If you want to know more about me, visit my website www.kathwalters.com.au or get in touch: kath@kathwalters.com.au

Kath Walters is a content marketing expert. She is the author of Australia's first book about content marketing, *Sticky Content: Mastering the Delicate Art of Content Marketing*.

She is the founder of a mentoring program for authors – 90 Days to your Brilliant Business Book. She also runs a training program for content marketers (bloggers) – Blog Bootcamp. Her clients include entrepreneurs, thought leaders, consultants, trainers and speakers.

As a business journalist and editor at Fairfax Media, Kath's articles reached up to 60,000 people each week.

CHAPTER TWO

THE 'SOCIAL CELL NETWORK'

HOW A NEW ONLINE MODEL OF
ENGAGEMENT, COLLABORATION AND
CO-CREATION OF SHARED VALUE
WILL CHANGE OUR WORLD

BY ALAN LLOYD

Founder cuuble: life sorted

I WAS INSPIRED TO RETHINK ONLINE ACCESS and usefulness by my mother's difficulties and my own desire to care for her from afar. I spent a lot of time with my mother in the last few years of her life. She was not IT literate. If she tried to go online, she was confronted with computer jargon, dozens of phone numbers and websites: one for the doctor, another for the dentist, another for the bank, and then all of her tradesmen. Every website wanted her to type in her email, personal details and create a password. She just looked at it and said, 'Oh dear' and gave up.

My mum didn't want to share her details with all these websites. She didn't trust them. She didn't even trust the tradesmen who came to fix things in her house. We all have trouble remembering phone numbers, names and passwords, let alone an elderly woman with a failing memory. My mum was sharp, but she couldn't remember all those connections and passwords.

Then there were the physical barriers for her. The broadband services in her retirement village were unreliable. Even with her glasses, she struggled to read most websites. She had a slight tremor in her hands that made using the keyboard and the mouse difficult. Why would she bother to access online organisations? The barriers were too many and too great.

Of course, you might argue that it doesn't matter much if the elderly can't access online products and services. But mum remained a consumer more or less until she died. She had money to spend. She wanted and needed a full range of products and services. She just couldn't get access to them with the way the web is set up at the moment with the multiplicity of web sites, screen designs, work flows and end user expectations.

ANOTHER WAY

I've been in tech all my working life. I knew the right technology could make her life a lot easier. So watching mum struggle with the net, really got me thinking.

> *Why, I wondered, couldn't my mum have her own web-services-based social network, populated by the people she knew and trusted, one that provided her access to all her service providers at the touch of a screen or with one simple click.*

With big buttons and typefaces, uniform accessibility options and a single-user interface with everything she needed, I could help her to do the crossword, or remind her to take her medications. She could phone me and my siblings with a click, answer our calls, share a picture, use a preset message and even turn the lights or television on and off from a single screen. A system that still had usable home-based functions when the internet or broadband services were inoperable.

I wanted her to have full control. If she had an argument with her sister, she could turn off her sister's access to her. If she didn't like a tradesman's work, she could remove him or her from her address book, and if he got back into her good books and she changed her mind, she could bring him back online – in a moment.

And, most of all, I didn't want her to have to find a whole lot of different websites, and log into them, and remember dozens of passwords, because I knew she wouldn't do that. Or, she would use one password for everything, or write her password down, and compromise her security. The final straw was when her medical centre wrote her a letter and advised that all appointments in future were to be done online!! I understand the efficiency opportunity for an overstretched and expensive medical system, but it came at a cost to my mother, just one of many people with accessibility difficulties.

OUR FRAGMENTED ONLINE WORLD

There are historical reasons why the online world is so fragmented and fast moving, and rushing past the people who arguably are in most need of it

There are technical reasons that explain why the online world evolved in this complex way – a model that served us well in the early stages of online retail, as one example. One reason is that the online shop is a model that we are all familiar with and understand. It is just like a real world shop; somewhere we visit to make a purchase. While we have been getting used to the web and integrating it into our lives, it is helpful to use old models that we know. It is easier to sell a 'better mousetrap' than a revolutionary concept.

Also we are at an early stage in the emerging power of the online world and not surprisingly it is an experimental, fast-moving and unresolved world. It is also an unfriendly world for those who have not grown up with it, who have sight, dexterity and memory problems. This is a large and significant part of the population, probably a significant market for products and services and certainly a group that online servicing could yield efficiency and cost reduction benefits. There are other reasons – that are not at all technical – why the web is so unfriendly for older and less able people. Technology departments are full of very young guys – mostly guys – who are unaware of the accessibility issues.

And the young like the whizz-bang new stuff. They are always searching for the new and different in technology, and they are trapped in a cycle of development that doesn't even consider the non-able bodied, or the intellectually disabled or the fact that many people have no or unreliable broadband services. They are not attuned to a new and different market, or a new and different business model. They don't really think about business models; they think about new technology. That is their job. And now, even the business heads in companies are starting to assume IT literacy amongst their customers. In a way, it's an old-world paradigm about what IT is: that IT people can't understand business and business people can't understand IT. IT people say the future is all about the cloud and big data and now the latest jargon is block chain. I know, because I've worked in IT R&D and IT business departments all my life. The new story needs to be about information, services, efficiency and profits.

As a result of the speed, excitement, discovery and ambition to discover and be rich from the next big app and disruptive business idea, the online world is increasingly complex and fragmented. Many aged, disabled and poorly sighted people experience this most acutely, losing the ability to interact easily with online businesses, social groups, care providers and eventually their family and friends. Isolation that can result is a catalyst for dementia and depression.

In terms of ease of use, consider:

- Fragmented functions, small screen devices and apps
- Multiple links, website designs, buttons and procedures
- Multiple registrations – terms and conditions
- Multiple logons and passwords
- Multiple - single process applications for medical and care applications
- Chaotic contact – address book references
- Duplicated and possibly dated information
- Even those with able minds and bodies have trouble with the complexity
- And those that don't may just not even try – and become isolated from services they need and the people they enjoy communicating with particularly a younger generation.

THE FUTURE MAKERS IN DIGITAL TECHNOLOGY

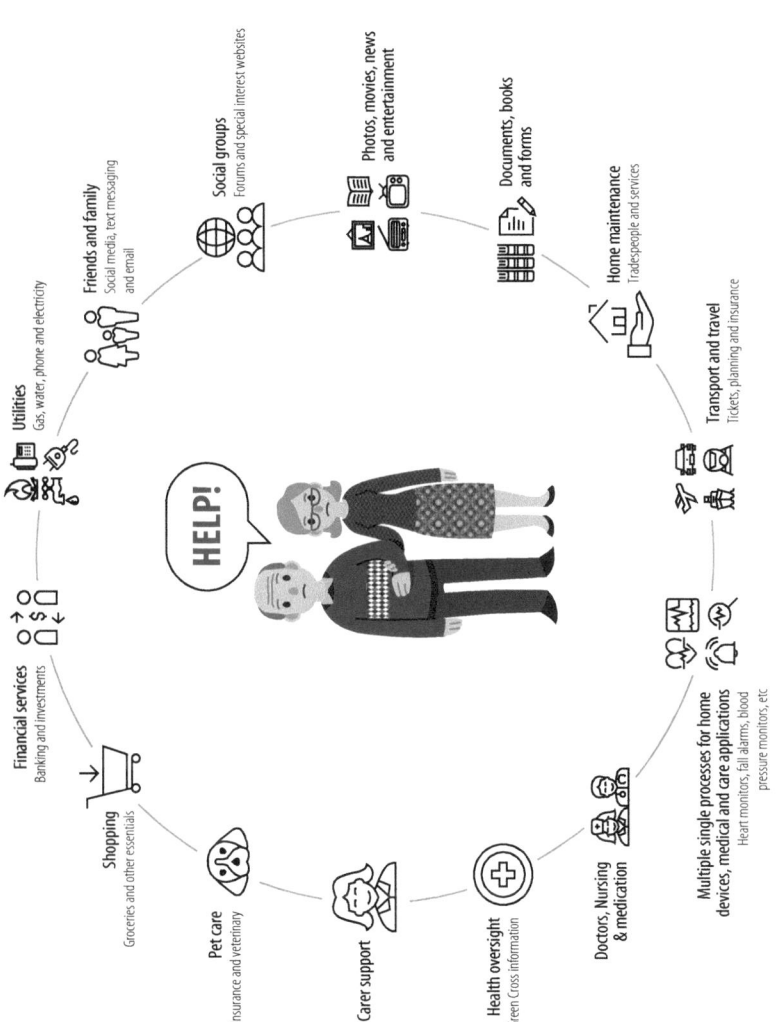

A NEW DIRECTION

I came up with a different way of making it easy for all people to gain the benefits of online access. I decided that everyone who wanted to deal with mum had to come to her. Here's how I thought it could work. I'd build a social network, or, what I call, a 'social cell network', and help Mum create her page and her world. Social cell is a term that I have borrowed from biology because it describes the smallest living entity in a system which can be made up of several or even millions of other collaborating and interconnecting cells – all of which share the same values to exist, interact and grow.

If you, other social cells or business systems wanted to engage with her, she would give you permission to connect with her.

That way, the companies that want to serve mum upgrade their systems to connect with her online. It is up to them. And they have an incentive to keep her world safe and usable, or get banned by her or a caring relative, immediately and dynamically.

I wanted to create the right of consumers to protect and control their own data, the basic principle being that data is an asset that belongs to the consumer and that privacy and management of your own identity is safer in your own hands than in the hands of another social network provider. Who better to keep your own information up-to-date than you, personally and who better to decide what information defines your needs and the value you are seeking from online suppliers?

I also recognised that simplification of the online world required a one-stop shop platform approach which can remain contemporary by being able to host new apps and functions but where a consistent, repeated approach, a personally managed image based address book and big buttons and letters gave ease of access to all users equally.

AND WOULDN'T THAT BE BETTER FOR US ALL?

It's not hard to see that this is an appealing business model for all online consumers. Imagine if we all had our own social network page – a personal social cell, interconnected if required with other social cells on a peer-to-peer basis to form groups, villages, care services, and where any company that wants to business has to knock on our door and ask us politely for access. We invite them in, and they do everything they can to stay in our good books. They start to understand how business, customer and care systems can work together.

I call it a personal social cell network. And I built it, with my business partner, Susan Oliver. It's called cuuble, and really, it is just the kind of social network that mum would have found really useful. As far as I know, cuuble is the only social cell platform in the world today.

And the social cell network will disrupt the old world models we have replicated in the online world.

THE SOCIAL CELL NETWORK MODEL IS DISRUPTIVE

It's fascinating to me that our online world is so 'old world'. Really. Online retail for example, is the exact same business model as the main street retail model: I build a shop and try to entice you to come to it. That is all a website is – an online version of a shop. But it is dumber than the high street shop, where the proprietor had some chance of knowing who personally came, who bought and who returned. It is also a questionable investment because we try to make these portals more appealing or more complex with forms and conditions, but they don't face the high street any more. They compete globally, possibly in micro-proportions in the hope that people will come. It offers an opportunity to make the online presence of an organisation a profit centre. For many companies, being online is an expense, but it's mandatory to have an online presence. For many it is no more than a billboard announcing their presence commercially and a signpost to how to contact them.

To other companies, it seems their web site is trial by fire: an administrative-process-defined minefield where the customer has to overcome the frustration-endurance test with forms and procedures, simply to become a customer.

And more recently, it is an open invitation for any advertiser to appear in your inbox with yet another unwanted and uninvited offer – a nuisance call in the era when email inboxes are 90% unwanted junk mail.

The customer is king. But just when companies are catching up and bowing to the will of the consumer by putting their products and services online, the consumer is getting more demanding about what they want, what they can deal with and the time they spend.

There's an increasing disenchantment with having to create passwords and profiles for so many sites. Companies want us to build our online profile so they can understand us and provide better products and services, but from the consumer point of view, it's tedious and gives away our information and privacy. Customers, particularly the ones with aging and disability issues, want lifestyle services that can include care, home deliveries, 'uber-style' services, home help, specific products and trusted service providers. They want business that is not just business, but socially sensitive business.

One challenge to making profits online is the difficulty of understanding the broader requirements of customers. The online world forces online retailers to work within a narrow band of understanding of what their customers want to buy and if they are satisfied with only that. The best they can do is use artificial intelligence to try to guess what else we might buy. The online bookshop, Amazon, is good at recommending books based on our searches or our purchases, or what other people who bought the same book went on to buy.

But Amazon sells a lot of other stuff, and it cannot find out whether we would like to buy it. It can't make a wild guess – people who bought this book like to go kayaking, for example. That sits outside of their algorithms.

But using the social cell network, we, the customer and individual, can share and interact as much as we want to with our

Business system interconnection with a meshed social cell system

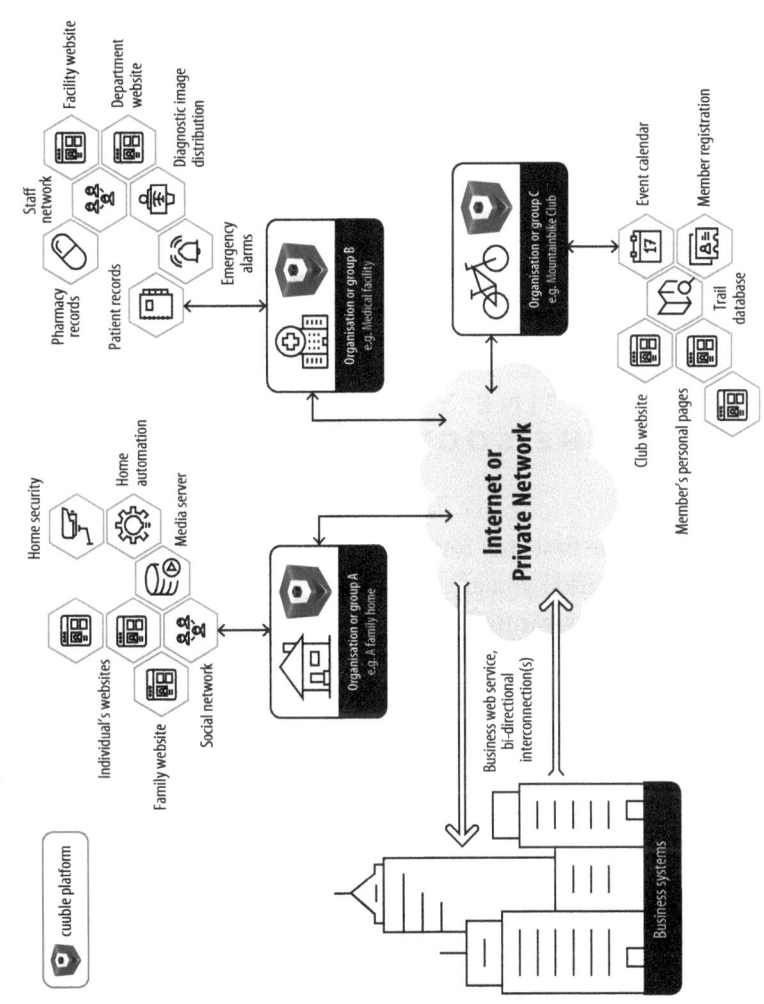

friends and companies. We can connect to other people and other people can relate to us as we choose. We can let our trusted social and business partners know what we want and need, and invite them to provide products and services and discounts and offers to entice us. And we do it once from our personally self-managed social cell.

Naturally as the number of other users and businesses engage in the social cell approach, each social cell system can interconnect with other social cells on a peer-to-peer basis to form a mesh type network.

This amounts to a new and more lucrative business model and greater business intelligence. I have written white papers on how this new model will help with business intelligence, the Internet of Things (IoT) and Identity management that you can download for free at *www.cuuble.com*.

RE-IMAGINE YOUR BUSINESS USING THE SOCIAL CELL MODEL

STEP ONE

You've taken the first step towards a new future if you recognise that there is a big market you are missing out on, and if you agree with me that the first movers into this market are going to uncover a new, more effective online and intimate business model on the web. In technology parlance – forget business-to-consumer. (Noting that centrally governed social networks are effectively B2C.)

But global social-business models need a lot more capabilities so they can connect using the social cell, and include identity and relationship engineering for the social-human interaction needs they embrace. And now you are no doubt wondering where to go from here. Of course, because you have invested in Business-to-Consumer and back-office systems for decades, it's a question of evolution and integration of these with the social cell network, where the social cell network is the new and emerging front end for your social-business customers. Even so, we have designed cuuble with a set of system administration functions and a set of operator oversight functions so that the branding and legacy system integration task can produce a basic in-house, operationally relevant prototype in a matter of weeks.

STEP TWO

Step Two is to decide where in this new map of e-commerce you would like to play. Personally, we've decided to play in the space of making this whole idea happen – creating the platform for change. A platform which you can own and run and integrate your legacy systems with. This platform is cuuble.

STEP THREE

A step into a model of engagement, collaboration and co-creation of shared value. Our social sell platform is 'white label' – some companies are going to choose to be the conduit between suppliers and customers and some are going to choose to participate as suppliers and co-creators, using the social cell network service model.

Alan Lloyd is an expert in making the online customer experience count. He works at board and operational levels to achieve information systems service delivery and technology change. Alan has worked with global companies and with governments for 45 years in 22 countries. Today, he creates social-centric care systems and integrates them with online businesses.

Alan is the co-founder and inventor of cuuble.com. This self-owned, self-branded and self-managed platform uses a distributed peer-to-peer mesh system for social and care-based networks. This allows traditional customer-centric systems to evolve into social-centric operations. The platform has its core functions certified to WCAG 2.0 – screen reader use for blind and those with sight difficulties.

Alan has followed the remarkable trajectory of modern technology through the past five decades. He has been at the forefront of large-scale convergence and identity engineering since the 1980s. From 1995 to 2008 he designed the wwiteware SDP/governance platform and other large scale online customer centric service delivery systems.

CHAPTER THREE

PITCH BIG

THE SAD TRUTH ABOUT WHY MOST PRODUCTS FAIL, AND WHAT TO DO ABOUT IT

BY SAM LANYON

Executive director and co-founder,
Planet Innovation

I NEVER WANT TO SEE ANOTHER GREAT product idea die because its inventor is misled and misguided in how to commercialise it. And even more importantly, I don't want to see the emotional and financial wreckage that our most inventive people – and their families – suffer when their great idea goes horribly wrong. Instead of reaching the market and making a contribution to the world, great ideas are lost and their originators are left jaded and bereft.

After decades working in product development and commercialisation,

> *I've seen enough failures to stoke a powerful desire to create a better ecosystem to support our inventors.*

And I know I am not alone. Throughout Australia's innovation community – investors, venture capitalists, business strategists, serial entrepreneurs and product developers (like me) – there's a growing sense that we are on the brink of having the smarts, the systems and the support to build a really flourishing community of product makers. But we need to debunk some of the most pervasive myths around product development.

This chapter is for all those who care about and work in the field of innovation, especially those companies and individuals who have a great idea and want to bring it to the world. Read it now, before you spend another dollar. It'll be the best investment you ever made. It contains all the secrets of success I have learned from my own experience of taking products to market. This applies to digital products as well as to physical products. And it applies to physical products with a digital element – too often left until the last minute and included as an add-on.

Ironically, the thriving startup culture that has emerged in Australia in the past 10 years has made it more important than ever that entrepreneurs get the right education about what it takes to bring a new product to market. Television program's like *The Shark Tank* – in which inventors pitch for investment from high-profile wealthy individuals – popularise the idea of inventiveness and entrepreneurship. That's great. But at the same time, these shows perpetuate many of the myths that put our entrepreneurial culture at risk. In this chapter, I want to bring those myths to light, and to outline a better way forward – a way that would supercharge our innovative culture, bring more ideas to light and greatly improve chances of new products successfully reaching the market.

The principles I'm explaining here are not theories – I have applied them again and again, successfully, as have with the other directors who co-founded our product innovation and commercialisation company, Planet Innovation.

PERFECTION IS NOT COMMERCIALLY SMART

We have to let go of the myth that the best way to get a product to market successfully is to focus on getting the product technically perfect. This is an old story, and it is so wrong.

An inventor approached Planet Innovation four years ago with a new version of a popular stationery product. There is already a big market for this product, and his innovation might have had a commercial future. But it never reached the market. We advised this man that we needed to build up an 'investment thesis' for his idea, and to test it with various markets.

He didn't want to hear that message. He came to us for advice on perfecting his product design. He thought he knew what perfection meant (although his market turned out to think otherwise). He thought he knew what his market wanted. We refused to take his money. But he found another company that was willing, invested another $2 million in his product. It failed.

I'm not sorry our company missed out on that $2 million in revenue because it is such a sad example of what happens when players in the innovation market perpetuate the myths of product commercialisation. Some of these myths really make my blood boil. Below is a list of some of those myths, and after that I am going to explain the realities. Just by the way, you might think that these myths are a thing of the past. They aren't. Every month, we see dozens of potential clients who are labouring under these falsehoods. And this is not the case just for small start-ups but even in the innovation departments of big companies.

So, here are some of the most deadly myths I have seen:

- Focus on perfecting the product.
- Stake everything you have on getting the design right.
- Be secretive about your design so that no-one steals your idea.
- Patent it nationally and internationally so you can't get ripped off by the large multinationals.
- Great products sell themselves.

THERE IS MORE MONEY THAN EVER BEFORE

The first step toward a commercially successful product is to get the 'investment thesis' right. The investment thesis is a clear, substantiated explanation of the market you are entering and what problems you intend to solve. Thesis is a good word for this document because it has to be based on a lot of research, just like a Masters thesis. Without the investment thesis, no company or individual can raise the capital needed to get a new product to its market.

In Australia today, it is both harder and easier to raise money than ever before. There are more investors in the business angel and venture capital markets who are willing to back great products. Of course, there is also competition for investment. Entrepreneurship is cool – in the 1980s, to be called an entrepreneur was like being labeled unemployed. Similarly, today there is an avalanche of information and advice – including misinformation and poor advice – that makes it hard for investors and entrepreneurs to get it right.

> *So why do so many inventors – individuals and companies start by researching their products and solving their technical challenges long before they approach a market, consider their business strategy and develop an investment thesis?*

Most inventors understand technology, but don't understand business or marketing. It's one of the reasons that they tend to stick to the lab bench or garage: it's comfortable and exciting there. Out in the real world of business and investment, there are 'sharks' who are ready to rip you off. Inside the world of your shed or lab bench is a creative and safe place. Progress seems real and tangible – a new version of a product or software. It's understandable that people hold onto this sense of safety, even sometimes to a tragic degree. Because by the time they emerge from the lab into the real world, there's not enough money (or energy) to get the product to market.

Even those entrepreneurs and companies who do understand the need for market research typically focus on market size, rather than how and why their customers will buy their product. For example, they will find out that the global stationery company, 3M, sells $1 billion worth of Post-it notes a year. Then they make a major leap in logic along these lines: 'If we capture just 5% of that market, our company will make annual revenue of $50 million a year.'

The problem is that persuading even 5% of a market to stop

buying a product they love and to start buying a new product takes a strategy. Imagine if I tried to persuade you that your favourite coffee shop was no longer the best coffee shop. Actually, a better example is that persuading people to change brands or products is more akin to trying to get them to believe that the world was not created through evolution. It's not easy. And, often, it is not possible.

THE ZEN OF MARKET FEEDBACK

Market feedback can completely change a product. Back in 2013, a company briefed Planet Innovation to produce a thermostat for home use. They told us they wanted all the bells and whistles on this thermostat. So we set about speaking to people in their potential market and to researchers. They told us they wanted the exact opposite – a thermostat with no bells and whistles. What people wanted was a simple control that allowed them to turn their heating or cooling at the press of the button. The only complex feature they wanted was the ability to use the thermostat remotely.

So we built our Zen thermostat based on an investment thesis that reflected this research: the market was full of thermostats that had a zillion different amazing functions. But the reality is no-one likes using them. Why? Because it is so hard to get the thermostat to turn on that, once they succeed, they are terrified of touching any other buttons. What people want is a simple on/off switch that allows them to control the temperature and turn on the cooling and heating remotely.

Cool and contrarian as that was, it was only our thesis about the product. Our next step was to to decide who might buy that thermostat for a price that made it worth it for us and for our clients. And we needed to know why they would buy it and prove that they would.

The internet of things

A key part of our thesis is that Zen is part of the internet-of-things trend – which is a trend toward all devices being connected to the internet. It's also part of a trend towards better energy management in the home.

We proved the demand for a simple thermostat for the home market by conducting a crowd-funding campaign for it with the main objective to assess the market interest and price sensitivity, not to fund the development. The response was overwhelming. Zen reached its modest $50,000 goal in less than 24 hours. So Zen could have been a single product – a home thermostat that was simple to use. Just do it, ship it, and then just sit back and watch the money roll in.

Instead, we made the decision to expand the horizons of the company. The best thing we did was we actually got someone into the business has 20 years in the utility and energy-efficiency sector.

We realised that the bigger, easier and less addressed market for Zen was small medium business. That the smart home was always going to have some head-winds around customer adoption and the cost of customer acquisition. But small businesses are always looking for ways to become more efficient, and their heating and cooling bills are big. That's what I'll call an investment thesis.

AND NOW FOR OUR BIGGEST SECRET: PITCH BIG

It's a sad fact that all that research is not enough. On it's own, even all that data will not convince investors to fund your company, a customer to buy your product, or a channel partner to distribute it. This next step is crucial on the path to commercial success. Yet it is the most misunderstood

It is this. 'Pitch Big'.

By Pitch Big, we mean develop your investment thesis into a 'story'. Craft a short, exciting and compelling story of how your product will change the world of your customers. That is what will engage the emotions of investors, customers and partners. Biotechnology companies are generally fantastic at pitching big due to the long development timeframes offset by the promise of their wonder drugs. They tell the story of patients who suffer awful diseases and how their products transform their lives. Charities are also masters of engaging us emotionally.

But Pitch Big doesn't mean your product has to change the world. It's enough to delight and inspire your audience with a little story about how your product works and why customers will buy it. You must support your story with the solid research in your investment thesis, but often, this story is what your audience will carry away to study later.

For example, I once watched an executive convince his board to fund a multi-million dollar enterprise software project with a straight-forward mock-up of the product's user interface. He spent a fraction of the budget on creating a bunch of screens that allowed the directors to experience what it would be like to use the real product. The board members walked out of that meeting saying the development was nearly done. And they were not joking because that was what they just saw. They saw and experienced life as it would be after the development was done.

PITCH BIG MITIGATES THIS COMMERCIAL RISK

We have run a dozen Pitch Big campaigns and every one of those has resulted in an immediate sense of clarity around the true commercial viability of the product. Sometimes, it is immediately clear that a product will fail. But whether you have a software app or a hardware device, if you can't raise money or secure interest from a sales channel, your enterprise will fail. That is the harsh reality.

As the previous example shows, it is possible to raise investment for a product well before you create a viable product. By this I mean that you could not actually sell the prototype you have created because you haven't solved all the design problems. What you do have, however, is an strong investment thesis and a terrific pitch.

At Planet Innovation, we recently raised $8 million in venture capital. Our investors are experienced people whose knowledge has been forged in the fires of failure as well as success. They can't be fooled. We had done the hard yards but at the time, we did not possess a tangible, market-ready product. We needed to employ the principles of Pitch Big to make sure that our vision and investment thesis made sense, and there was enough substance to convince our investors that the reward well outstripped the risk.

IMAGINE SUCCESS

Imagine the life that follows the commercial success of your product – it's rewarding, healthy and happy too. Imagine that you think of an idea, test the market, build and investment thesis and then Pitch Big. As Dorothy says to her prodigal husband, Jerry Maguire, in the movie of the same name: 'You had me at hello'. If you Pitch Big, you'll have them at hello.

Imagine this:

- Instead of spending $1 million on developing your product, you spend a tenth of that. Then you look for and find investment.
- You are not starving; you are thriving.
- You are not alone; you are surrounded by people who believe in your vision.
- You are not bitter and disappointed; you are excited and happy.
- You are not broke; you are on your way to commercial success.
- You are fulfilled: you have left your legacy.
- You've helped to make the world a better place.
- You don't try to do everything yourself, you are surrounded by talented people who help get where you want to go.

START NOW ON THE PATH TO COMMERCIAL SUCCESS

1. **Stop now**, if you are building a product
2. Create your investment thesis – a clear, substantiated explanation of the market you are entering and why, based on research, just like a PhD thesis.
3. Pitch Big. You can start by building a powerful but simple website. Then create a compelling, impressive animation, a persuasive powerpoint pitch deck and a brilliant brochure. Get out there and talk to prospective customers, channel partners and investors.

Easy, eh? Not quite. Starting from scratch and taking a product to market isn't easy. Here's are the challenges that might derail you:

- Long hours – while your work will be more purposeful, it's still hard work with long hours. Pace yourself, and work strategically.

- The wrong advice: everyone will have some advice, many of them will be people who should know better than they do – government officials, inexperienced investors, friends who have never commercialised a product but have great ideas about how to commercialise yours.

- Neglecting your closest relationships with your loved ones. Don't. Keep the communication open, plan breaks and work in 'sprints', with periods of lower intensity.

- The comfort of the lab bench. We understand inventors. We are inventors. We know how much you love to tinker because we love it, too. But this is the number one cause of failure that we see – too much development without enough market research. If you stay in the comfort of the lab bench, tackling the gnarly but satisfying problems of development instead of working on annoying stuff like your investment thesis, your story and your pitch, you might succeed … but you might be another statistic. I don't want that. That is why I've written this chapter.

MY WISH FOR YOU: MAY YOUR PRODUCT SUCCEED AND CHANGE THE WORLD

I want you and everyone in our innovation community to know what I and my fellow directors at Planet Innovation have learned from taking real products from the lab bench to world markets.

I invite you to contact me

If your product idea is just a twinkle in your eye, so much the better. If you've poured your money, heart and times into it, come with an open mind. Here's a link to prepare for our meeting www.planetinnovation.com.au/about/our-principles/

Sam Lanyon is the co-CEO and co-founder of Planet Innovation. He is an innovative and strategic business leader with significant experience in strategy, sales and operations with a successful track record in the global commercialization of technology-rich products. Previously, Sam was the International Sales Director for Vision BioSystems where he was responsible for establishing the sales, marketing, customer support, and service operation throughout the EU, Middle East, Latin America and Asia Pacific.

Sam holds an Honours degree in Mechanical Engineering from the University of Melbourne, a Post Graduate Diploma in Management from Melbourne Business School and strategy training from London Business School.

CHAPTER FOUR

TOWARDS PURPOSEFUL COMPANIES

HOW TECHNOLOGY SHIFTS THE PLAYING FIELD

BY TOM DAWKINS

Chief executive and founder, StartSomeGood

THE 'PURPOSE ECONOMY' IS HERE. Right now, it is unevenly distributed and poorly understood, as the future often is. But in the coming year, the purpose economy will redefine how we conduct business and derive meaning in our lives.

If you are a company leader and want to avoid newer, more socially-minded startups superseding you, this chapter is for you. If you are an entrepreneur who intends to put those old companies out of business, this is also for you. Each of you has an incredible opportunity.

We are at a similar moment for the purpose economy as we were with the internet economy 20 years ago. Most don't yet realise the impact to come. Those who do are already deriving significant tactical advantages. Soon they will become huge strategic advantages. And then they will emerge into the new business-as-usual.

We saw one of the early signs of this shift in 2014 when TOMS Shoes, an eight-year-old company that gives a pair of shoes away for every pair it sells, sold 50% of its equity to Bain Capital, a private equity investor, for over $300 million USD. At the time, some worried that Bain could end footwear donations, but they declared they would do no such thing. So far, they have been true to their word. Private equity investors are not known for their sense of social mission, so what's going on here? The answer is that Bain understood that TOMS' mission is not a cost; it is an asset. It is the core reason they had grown to a $650 million company in just eight years.

TOMS is one example of a growing part of the economy in which businesses compete not just to create private wealth but social outcomes. These are 'social enterprises': companies that have reaped huge rewards from being more purposeful than their competitors. They include The Body Shop, Ben & Jerry's Ice-cream, Credo Mobile and Kickstarter, as well as an emerging wave of startups in every category imaginable, some of which you'll meet in this chapter. In Australia, where I am based, we see the success of TOM Organic (no relation to TOMS Shoes), Five:am Organic, Future Super and Thankyou Group amongst many others.

The world is changing before our eyes, for those willing to look. Millennials are demanding a new level of meaning – of purpose – in the companies they work for, buy from and invest in. This change is the big challenge for business-as-usual and creates enormous opportunities for those who get ahead of this trend.

What has this got to do with digital technology you might ask? Everything. Technology is the enabling environment that is bringing the purpose economy into being. It is the toolset used by smaller social enterprises as they out-innovate and out-compete established businesses. New technologies allow us to transmit and share our values in a new way. And our newfound interconnectedness changes our values. Consumers, empowered by technology, are pulling back the veil on destructive company practices and supporting businesses that are a force for good. Whereas being an ethical consumer was once a laborious chore new tools are making it easier and easier.

> *This is a story of changes wrought by technology, not about changes to technology. The effects of these changes will be more all-encompassing than most realise, and if you want to ride this wave you need to know what's going on.*

I see this trend every day in my role as the chief executive officer of StartSomeGood.com, a crowdfunding platform I co-founded in

2010, which has helped launch over 750 social enterprises at the time of publication. Before StartSomeGood, I spent two decades working with young changemakers and entrepreneurs, both in Australia and overseas. I've founded non-profits and social enterprises, worked for a global non-government organisation in Washington DC and for a startup in San Francisco.

I've been building social impact projects since I was 16. By the time I graduated university, I had launched three non-profits, each of them aimed at inspiring young people, my peers, to be active citizens. One of those organisations, Vibewire, is now 17 years-old. During the eight years I ran Vibewire, we published several anthologies, hosted a film festival that toured our nation, sent youth reporters on the campaign trail with the political leaders of both major parties during the federal election and opened the first co-working space in Australia.

Along the way I ran a dance music event company, was employee number four at Australia's most successful youth culture website and founded the Australian Changemakers Festival. I was the first social media director at the influential international non government organisation, Ashoka: Innovators for the Public.

Over the past two decades, I've helped hundreds of projects launch. Here's what I have learned: there's never been a better time in human history to get a project off the ground. The barriers to entry have come down in every way conceivable. The cost of hosting: from running your own server like we did at Vibewire to the ease and scalability of Amazon Web Services today. The challenge of managing technology: access to scalable, easy-to-use platforms to help you build websites, handle email lists and run fundraising campaigns. The ease of recruitment and collaboration across countries and time zones.

Technology has shifted the playing field for new companies in three important ways, all of which are to the advantage of purposeful businesses or disadvantage those without purpose. Those three ways are:

1. Greater awareness of options
2. Increased cultural sharing
3. Reduced barriers to new competitions.

GREATER AWARENESS OF OPTIONS

There has always been a group of ethical consumers. Estimates of their numbers vary, but market researchers have pegged them as about 15% of the market in advanced western economies. The problem wasn't that more people didn't care. The problem was that it was too hard, or too expensive, to prioritise ethics while shopping. The way to be certain was to purchase direct from boutique companies that specialised in this niche. For example, near where I live in Sydney is a small all-organic grocery store. Ethical consumption used to mean going out of your way and spending a bit more than the mainstream equivalents.

But ethical consumption in the future will look different. Just look at egg-purchasing habits. I no longer need to go to boutique stores to find free range eggs. In fact, Coles, Australia's second biggest grocery store chain, no longer sells caged eggs. Free range eggs have become over the mainstream. And once purposeful products are right there on the same shelves, equally convenient and close in price to the less ethical product, they will take over more and more mainstream categories.

A sustained online and offline campaign shifted egg-buying habits, driven by raised awareness of the impacts of this choice.

In the same way, awareness is now growing of the existence of better options in so many categories as the idea of social business becomes more celebrated and better understood.

For most products, however, the choices are not as black-and-white as free range versus caged eggs. The issue becomes not just will (the desire to do the right thing) but information (how to choose the most ethical option). To support more purposeful companies, we need to know which is which. And, once they empowered with this information, a much higher proportion of consumers are making the pro-social choice.

Peer endorsement

In a survey conducted by market research company Core Communications, 71% of consumers agreed they would be more likely to purchase from a company that supports a cause they care about. This number is growing as the ability to differentiate those doing the right thing becomes easier and easier, thanks to technology.

How does our growing awareness occur? Via our friends' endorsements (which I cover in detail below), and via tools, platforms and labelling systems that make it easier to identify products and companies who are congruent with our values.

This large reduction in the barriers to and costs of conscious consumption has increased the number of those who shop with awareness. With social media, many ethical products are just a click away and, as in the case of crowdfunding supporters, most new customers come via personal recommendation from a peer.

Third party endorsement

But even if we're out and about without a specific brand in mind, navigating the shelves of mainstream stores to find the more ethical products has never been easier. There are an increasing number of third-party certifiers like Fair Trade and B Labs to help us identify the best products based on their criteria. As these products continue to infiltrate mainstream retail they grab our attention as we are making our purchasing decision.

StartSomeGood became a B Corp in early 2016, the 90th in Australia and 1500th worldwide. B Corps are a growing movement of businesses committed to making an impact. To be certified you must submit to a rigorous and comprehensive evaluation of your business activities, from salaries to suppliers, leave policy to volunteering. The effort is worth it because B Corp certification is responding to a demand from consumers for help identifying the most ethical products, and being identified as one of the good guys has real business value.

Today we have access to tools like Good On You, an app with which you can scan barcodes of clothes to find out about how they are made. Or Folo, a browser plugin which alerts and guides you to the companies and products giving a share of their profits to social causes.

Changing employee preferences

Employees are choosing purpose too. Certifications like B Corp also helps many talented employees decide who they want to work for. Platforms like Glassdoor are allowing employees to anonymously rate their companies, increasing the transparency into a company's culture and values. As a business leader, this matters to you. What is important to many of the most talented people you will want to employ is more focus on meaning and impact, especially amongst millennial workers entering the workforce.

A recent survey from the consulting firm, Deloitte, shows that 72% of employed Americans say they would prefer to work for a company that supports charitable causes when choosing between two jobs that offer the same location, pay and other benefits. For millennials, these numbers are even stronger. In a 2013 survey, 93% of millennial workers said that they expected businesses to have a social commitment; a mere 7% agreed that the only job of business is to make a profit.

In this way, millennials are forcing us to re-evaluate how business works. They don't want to accept the old binary of either shutting up and focusing on earning money in the corporate world or spending your life in poverty trying to make a difference.

Instead, they want to do both together, and so they should. We should be able to make a living, and even a healthy living, while also doing work that creates a positive impact. Companies that provide that sense of meaning, who have a purpose that goes beyond profits, have a real advantage when it comes to recruitment and retention.

A commitment to purpose, therefore, is providing measurable business benefits today, not just at some theoretical point in the future.

INCREASED CULTURAL SHARING

Many companies have embraced social media as a new low-cost marketing channel. But it's so much more than that. Social media doesn't just allow you to get your message out; it reveals who you are. Social technologies have changed the playing field for social enterprises, enabling them to grow faster not through big marketing budgets and the interruption-based tactics of old but the passionate advocacy of their community.

Social media is the greatest tool for sharing that has ever existed, but that's not much good if your story isn't worth sharing. Social enterprises are more worth talking about than regular products. In sharing the fact that you have purchased from a social enterprise, you are sharing something about yourself: your values and beliefs.

People buy stories

As Blake Mycoskie, founder of TOMS Shoes says 'people buy stories, not products.' Something Blake discovered was that his customers would sell his product for him, not because it was the most stylish or best value (although it does well on these dimensions) but because it resonated with their values and gave them the opportunity to share those values.

Here's how Blake describes getting the first boutique to stock TOMS in his book, Start Something That Matters: "I went in and told her [the buyer] the TOMS story. Every month this woman saw, and judged, more shoes that you could imagine. More shoes that American Rag [the shop] could ever possible stock. But from the beginning, she realised that TOMS was more than just a shoe, it was a story. And the buyer loved the story as much as the shoe, and knew she could sell both of them."

A purpose for a business that goes beyond profit and helps creates the future we want is a different and more engaging story.

People also buy hope. And in a world beset by significant challenges, from climate change to increasing xenophobia and inequality, we burn to do something practical to nudge the world in the direction we want it to head.

Most social change is created through infinite cumulative nudges like this, punctuated with dramatic moments.

Social enterprise is a story both made up of millions of small steps – choosing the right product over the bad one – but it also represents a transformational step – the emergence of a new type of capitalism.

The combination of these elements – practical small-scale effects from each specific purchasing decision AND a pathway towards the big-picture changes we need – make every social enterprise product so much more than that product. It turns a product into a story, an individual act into an opportunity to express something fundamental about yourself. It turns the ordinary into the extraordinary, the mundane into the shareable.

And being extraordinary, being sharable, is the key to succeeding in the new world of social technologies.

Consumer participation in spreading your message is marketing Jiu Jitsu. Instead of using brute force to win a fight - or spread your marketing message – you leverage and focus the strengths of others to create your impact. And if you're not extraordinary in some way, if you don't have a story that engages and inspires, you will be ignored. As the marketing guru Seth Godin said in his book *Purple Cow: Transform Your Business by Being Remarkable*, 'In a crowded marketplace, fitting in is failing. In a busy marketplace, not standing out is the same as being invisible.' There is no marketplace more active and more crowded than the market for attention today.

The new marketing that social technologies require is about engaging your customers and inspiring them to share your message. For someone to share your message, your message must be personal to them. Otherwise, they might consume your product, but they won't broadcast that fact. Your product must say something about who they are and what they believe. In this environment, it's stories around meaning that stand out and get shared, as people look to assert and spread their values.

We see this a lot with crowdfunding. One thing you always get when you help crowdfund a project is a great story. I bought underwear through a company, Mighty Good Undies, that recently launched via crowdfunding on StartSomeGood. I'm not one to talk about the act of purchasing underwear. But I must have told a dozen people about Mighty Good underwear who launched in the last couple of months because their mission to make the world's most ethical underwear is an interesting story. Much more interesting and less creepy than, 'Guess what? I have new underwear!' And here I am talking about them again.

TOMS Shoes founder Blake Mycoskie describes the big realisation he and his team had when they started selling their buy-one, give-one shoes: 'People who tell the TOMS story are more than just our customers, they're our supporters. People who buy TOMS like to talk about their support for our mission, rather than telling people they bought a cute shoe from some random shoe company. They support the story – and the product – in a way a casual shoe-buyer never will.'

And it's not just that people will share your story, they will often be willing to pay more for your product because the purchase means more to them.

Many companies create outdoor jackets, but the giant sportswear retailer Patagonia has built a powerful brand based on their commitment to 'use business to address the environmental crisis.' For Patagonia, that means trying to get their customers to buy fewer jackets because the more they make, the greater the negative impacts. So, they focus on using the highest-quality materials, making and distributing their clothes in the most environmentally conscious way, with a commitment to quality backed up by a lifetime warranty.

While Patagonia sells their jackets for more than their competitors, this isn't an apples-to-apples comparison because Patagonia isn't just selling jackets. They're selling a story of environmental stewardship and inviting you to be part of that story.

As founder Yvon Chouinard puts it in his autobiography Let My People Go Surfing: The Education of a Reluctant Businessman, 'my first principle is that selling ourselves and our philosophy is as important as selling a product. Telling the Patagonia story and educating the Patagonia customer … on environmental issues is as much the mission as selling products.'

You are part of a bigger story

When it comes to marketing today, you have two choices. You can spend a lot of money. Or you can have a story that people will share for you. If you're a new company, option A is not an option (and is no guarantee of success anyway).

Instead, create something that people will market for you. Now more than ever they have the tools to do so effectively, at scale. But tools on their own don't go very far. They've got to have the motivation to share your story as well. That comes down to who you are as a business, what you're creating, and what that means to them.

Mighty Good Undies put their call to action this way: 'How you vote with your money and where you choose to buy is powerful – you are part of a bigger story about making the fashion world more ethical and sustainable.'

Social enterprises are emerging in almost every category you can think of, and all of them have this same meta-story. They invite you to be part of this bigger story of making business more ethical and sustainable. We've seen huge successes in shoes, TOMS; clothes, Patagonia; eyewear, Warby Parker; ice cream, Ben & Jerry's; cleaning products, Seventh Generation; and ecommerce, Etsy, Kickstarter. The passionate and motivated sharing of the larger story by their customers fuels their success.

As New York University professor, sociologist Clay Shirky has put it, 'Now that media is increasingly social and innovation can happen anywhere, people can take for granted the idea that we're all in this together.'

We've already dealt with the first and last part of that idea. Firstly, social technologies are binding us together in new ways, increasing our ability to influence those around us with our values. And a growing group of consumers give greater importance to the way goods are made, based on this idea that 'we're all in this together'. The positive impact their choice of product can make matters to them and guides their selection of companies to support. So now let's now turn our attention to the middle part – that innovation can now come from anywhere.

REDUCED BARRIERS TO NEW COMPETITORS

When I founded the youth non-profit Vibewire in 2000, it was before what we now know as social media. That was the age of online forums when most people weren't active online and anonymity was the rule for most that were. The reason the first question people asked when you entered a new forum was 'a/s/l?' – as in age/sex/location – was because we had no idea who or where you were. There was no Twitter, no Facebook (in Australia) yet, no geo-targeting.

It was hard to reach out to people with shared interests in other places.

But my co-founders and I were determined that Vibewire wasn't going to start life as a Sydney thing. We wanted to be national from day one, meaning we wanted co-founders from Melbourne. We figured the place to start was to look for people like us: university students with political interests and time on their hands who wanted to gain experience in entrepreneurship, journalism and project management. But how to reach them?

We realised there was nothing for us to do other than to go to Melbourne and stick up posters in the places they frequented: the three major university campuses and some nearby cafes and bars. A friend and I took a day off and took the overnight train to Melbourne, ran around sticking up posters all day before catching the overnight train back and returned to university the next day. I'm not even sure we had time to shower.

Compare that difficult, expensive (and smelly) experience to the launch of StartSomeGood in 2012. Then, my co-founder Alex and I didn't just want to be national from day one; we wanted to be international. Our challenge was finding projects to launch via crowdfunding whose founders both hadn't yet chosen a platform and would be willing to take a punt on something they couldn't even see yet: a new platform that would launch with them. How to reach them?

Through our networks of course! This time we spread the word via Facebook, Twitter, LinkedIn and personal and professional networks. With the help of endorsements from key influencers and friends inviting friends we managed to find 12 projects from 10 cities in four countries, and none of them came from where Alex and I lived, Washington DC and San Francisco. That's the power of interconnectedness.

Zero-cost startups are your competition

As Clay Shirky said, innovation can come from anywhere, and the cost of pursuing it is trending towards zero.

So what does that mean for existing businesses? It means you're going to get wave after wave of new competitors. That's how capitalism works, of course, but the threat is of a new scale and velocity today.

You'll have more nimble companies exploring new business models, including those that remove you from the equation. And you'll see new competitors offering your products and services in a more ethical and purposeful way.

> **As the cost of launching businesses is falling more people are stepping into the space between traditional business and philanthropy and creating companies that pursue a social purpose.**

And if you're not careful they're going to out-purpose you, making their story more shareable, their customers more passionate, their recruitment more successful, their business, in a word, more competitive than yours.

Crowdfunding is an early-warning system for this trend. It is here that we see the future of commerce first. What sorts of businesses and products are emerging as barriers fall, and which projects do people want to support?

Before crowdfunding one of the principal barriers to new companies in any product industry was the challenge of financing up-front. You needed to make the things before you sold them, putting many entrepreneurs into debt, making others reliant on outside investment and excluding many who couldn't access either path.

Now there's another, better path. Sell your products first via crowdfunding, then use these funds to manufacture, so you know people want it, and you avoid financial ruin if they don't. In this way entrepreneurs reduce their personal exposure to failure and can launch new businesses in a faster, leaner way. All of which is accelerating the rapid growth of social enterprises.

At StartSomeGood, we've helped launch new ethical companies selling food, including Wandu, Stories Food Truck, Eat Me Chutney and Harvest Fair; clothes, including Mighty Good Undies and One Night Stand; stationary with Words with Heart; 3D printing for orthotics with AbilityMate; co-working with Catalyst Collaborative; media with IndigenousX; apps including RipeNearMe, The Darwin Challenge and Open Food Network and even beer with the Good Beer Company. And this is not to mention soap, education, eco-tourism and subscription boxes, all of which still barely scratches the surface of the diversity we see. If you don't yet have a social enterprise competitor, you soon will, because someone will be passionate enough, and the costs are low enough, that they'll soon try. Mighty Good Undies, for example, was inspired by the #WhoMakesOurClothes social media campaign after a garment factory fire in Dhaka, Bangladesh, claimed the lives of 117 workers in 2012. Dissatisfied with the available options, co-founders Hannah Parris and Elena Antoniou decided to create the world's most environmentally-responsible and good-for-workers' underwear brand. Rather than build an expensive e-commerce website they tested their concept with a crowdfunding campaign, selling almost $40,000 worth of underwear on StartSomeGood to kick off their enterprise. They launched at Berlin Fashion Week in July 2016.

From niche to mainstream

Operate in business-as-usual mode, ignoring community interests and focusing on profits and you'll soon find yourself up against entrepreneurs like Yvon, Blake, Hannah and Elena. If you're not careful, you will discover that you're on the wrong side of history, outfooted by competitors born in this new purpose-driven environment, just as traditional businesses have struggled with the rise of the web and mobile-native competitors.

As technology and purpose continue to develop hand-in-hand, even the biggest companies are vulnerable to disruption.

Take the big four banks in Australia, that are now investing in 'FinTech' (Financial Technology) startups after being slow to the game. They realised that if they don't invent the future of banking, other companies will create it for them and that future might not involve banks at all.

Just as crowdfunding allows social entrepreneurs to bypass venture capitalists and philanthropic foundations and raise funds from their community, so too can peer-to-peer platforms makes it possible to skip the banks and lend to each other. And this is what's happening. The global market for peer-to-peer lending was US$26.16 billion in 2015 and is expected to reach a staggering US$897.85 billion by 2024, Transparency Market Research reports.

Banks once served a local community and created real relationships within that community. People knew the bank manager. The bank manager cared about their small business and could cut them a break if times were tough.

That sense of context and intimacy has been lost, as banks have focused on efficiency, scale and profitability. Banks are now among the most profitable companies in Australia, so they're good at extracting the maximum possible wealth from their activities. But they're failing to serve people who are unable to access loans for their businesses, or who are being squeezed by exorbitant credit cards or are struggling to compare loan options in an opaque market.

Technology is allowing a new group of companies to respond to these challenges, from allowing individuals to lend to each other to making comparison shopping easier to replacing banking infrastructure with telecommunications infrastructure in Africa, where people are using phone credit to transfer and store money, bypassing the banking system entirely.

Technology is re-balancing the human and community scale, allowing a return to older forms of collaboration and trust which has been atomised by big business as in banking. When I spoke to the author Rachel Botsman, who coined the term 'collaborative consumption' in her book *What's Mine Is Yours: The Rise of Collaborative Consumption*, she got excited about this point. 'That's the beauty and power of the idea [of collaborative consumption]!' she said. 'It is based on old market behaviours – swapping, sharing, lending, renting, bartering – reinvented through technology. The idea of collaborating and sharing is innate to us. We traded and exchanged in villages and market squares based on our reputations for thousands of years. Instinctively we know how to collaborate and trust "strangers"; technology is accelerating and scaling the ways in which we can do so.'

Even in the driest and most private-benefit-based markets, such as superannuation, there are purpose-driven competitors. In Australia, if you wanted to invest your superannuation in a focused, ethical way, you had to set up a self-managed super fund. That is expensive, time-consuming and too complicated for most people. Now, you can choose to go with a values-led investment fund like Future Super, Good Super or Australian Ethical Investments. If you want to bank in a way that supports social outcomes, you can join act., a new service from the Australian bank, Community Sector Banking.

These investment companies aren't asking you to sacrifice to do the right thing. They offer a product they claim is every bit as good, if not better than their competitors, and often at a competitive price, alongside a more positive environmental impact. It's no longer either/or.

It's this combination – at least as good in quality and value, and with a better impact – that is driving the growth of social enterprises. The market is no longer a small niche of consumers willing to pay more or go out of their way to do the right thing; it's a much larger cohort ready to make the right choice when given the option. And with transparency and peer-to-peer sharing increasing access to new opportunities and new social enterprises launching to fill existing gaps for social products, this is soon a choice all of us will have each time we shop. Once that happens, what was a niche will be mainstream.

It's more than a marketing message; it's about deep change

What I've covered so far are tactical advantages enjoyed by more purposeful organisations, made possible by the lower costs, transparency and sharing that technology offers.

These tactical advantages – in recruitment, recognition and marketing – are real and are important. Tactics after all are how you win battles. But strategy is how you win the war. Soon social enterprises will be more desirable to work for, more able to rely on the advocacy of their consumers and more dynamic in their use of new technology. They will have profound systemic strategic advantages that will be impossible to compete with from the business-as-usual standpoint.

At some point most countries are going to put a price on carbon and work to surface the externalities hidden by our economic model. True-costs economics, when we factor the environmental and social impacts of products into their price, is the inevitable consequence of living on a finite planet. You can still use resources in a wasteful way; you just need to pay for them. As climate change impacts increase, this shift is inevitable.

And when that happens, when those previously-ignored impacts like carbon emissions are priced in, the companies that have costed those externalities already, that have built their business with the reduction of those costs as a goal, are going to dominate. They're going to be set up to win this new business world.

As Patagonia's Chouinard says 'When we moved to organic cotton in all our cotton products 20 years ago we took a severe financial risk. But today the business and environmental impacts of our decision have been enormous. We know how difficult it can be to solve environmental issues and we believe that commitment to do so will lead to financial success. We've seen it time and time again at Patagonia.'

WHAT THIS MEANS FOR YOU

You want to be successful in the emerging purpose economy, so design a business that can win in this future. The shift in values is already happening, with millennials revealing a different attitude towards work and consumer interests that were niche becoming ever more widespread.

Social enterprises are succeeding in every industry, fueled by their greater adaptation to the world of social marketing and digital commerce. To thrive in the future, you won't be able just to market yourself in the same old way. You will need a story that resonates, a story that your customers are prepared to share with their friends and family. You need to be the kind of company people are proud to support.

The businesses that fail to adapt will be those stuck in the past, seeing social good as a cost rather than a strategic asset that can be leveraged to create greater business success.

When your customers know you do what you do for the right reasons – rather than just extracting the maximum amount of profit out of them – they're going to stick with you, and even collaborate with you to fix problems in your business. They are going to share your story and endorse you to their family and friends, who trust them much more than they trust you. The best young staff will come and work for you and will stay longer and work harder. All those factors make your business more successful. Those aren't costs; those are assets.

AN INVITATION TO CHANGE

If you want to join those of us already on the road toward sustainable, purpose-driven business models, I'd love to hear from you . If you want to explore how we can work together contact us via www.startsomegood.com/parntnership. To connect with me personally, or have me come and speak at your organisation or event, please contact me at www.tomdawkins.com.au

Tom has explored technology and culture for over 20 years. He specialises in using technology to create a more democratic and participatory society.

Tom is co-founder and CEO of the crowdfunding platform StartSomeGood.com, which supports social entrepreneurs to raise the funds they need to make a difference. So far, StartSomeGood.com has raised over $11 million for social impact projects in 30 countries.

Tom has founded several non-profits. He started Vibewire Youth Inc in 2000 while he was at university and lead it until 2008. Opening Australia's first co-working space was among the company's innovations.

Tom was the first social media director at US company Ashoka: Innovators for the Public. He took Ashoka's Twitter following from zero to over 300,000.

He was the Director of the Australian Changemakers Festival 2013 and 2014 and has been a regular speaker at conferences, seminars and workshops for past 14 years.

CHAPTER FIVE

DIRECTING CHANGE

THE STRATEGIC ROLE OF THE DIRECTOR IN THE DIGITAL WORLD

BY SUSAN OLIVER

Independent executive director

PROFESSOR JOHN McGEE formerly of Templeton College, Oxford, asks us as strategists to never believe there is such a thing as accepted industry practice:

> '... in the end, your most potent competitive weapon as a strategist will be your ability to systematically challenge the prevailing assumptions in your firm or industry.'

When I was a British Council Scholar at Templeton College, Oxford, Professor McGee introduced me to the concept of the 'old game' and 'new game' strategies. Since the second world war, countless research papers, books and *Harvard Business Review* articles have prompted organisations to systematically analyse their business, and use strategy tools to create winning business approaches and new models. It was never easy, not always well done and never a static, one-size-fits-all process.

Over that period, we saw the strategy approaches change. They ceased to be a matter of 'seize that strategic position' against a known enemy using known competitive weapons. Instead, they became guerrilla skirmishes over shorter and shorter time periods. Innovation came and went. It was often marginal to the business rather than a fundamental rethink. That is the world we grew up in, and the expectations and strategy processes we undertook in our boardroom and senior management roles.

> *But the role of the strategist is now being undermined. The participation of boards of directors is no longer clearly adding value.*

The problem for our businesses today is that our challenges do not fit into orderly strategy processes; they spring from an exciting, energetic hodge-podge of newcomers to the business world. They see a different way of doing things and have a go at it. They pick off parts of the value chain, thumb their noses at regulatory and geographical barriers and apply global models to suit a global citizen ready for the change. These new comers disregard traditional governance regimes, geo-political borders, and national tax regulations in some cases. They develop businesses outside the interests of any one nation or organisation. They disrupt our old models and companies are racing to keep up.

With all of this happening, the importance of Professor McGee's new game strategy approach takes on new importance. Until recently we agreed our old game plans were sufficient, executed better than anybody else, achieving lower costs or niche market positions -- although that was never my preferred position on strategy. Informed by Professor McGee, I pushed board colleagues to think outside of the square, to develop creative strategies – 'new game' strategies. In some cases, we did move to the new game, but mostly business-as-usual won the day.

I propose that we no longer have an option on new versus old game strategy. Time has run out. Our hand is now being forced by the sheer quantity and imagination of disruptors coming from unexpected quarters. We must seek new game strategies. The innovators are on a roll and amassing globally! Nothing is slowing down and waiting for us to catch up. They address the youth markets and shift the way our society sees its national boundaries and the business organisations in it, and even the way our governments govern.

The challenge to boards is this: how do we understand where the disruption is coming from? Can we rummage amongst the hype, the talk of unicorns and the sheer multiplicity of start-ups, apps and well-funded, new and fast-growing technology developments to identify the threats and opportunities relevant to our businesses and society and their future? There seems to be no choice but for senior managers and directors to understand this landscape as much as possible. We must set ourselves the task of acquiring digital intelligence and figure out what is the extent and reality of the digital challenge.

AMAZONED OR UBERISED

You can see it as alarmist or a call to action – every industry is facing disruption. As Pierre Nanterme, the chief executive of consulting company Accenture puts it: '… new digital business models are the principal reason why just over half of the names of companies on the Fortune 500 have disappeared since the year 2000.'

Amazon's trajectory from bright idea to global domination is a leading example. Amazon, started by Jeff Bezos in 1994, has transformed and dominated the book publishing industry. It now commands 67% of the book industry, while every other company combined, from book retailers to publishers, gets 33%. Amazon model overturned the publishing industry with the introduction of e-books. The publishing industry just stood aside and let Amazon do the innovating. Yes, book publishers and retailers saved themselves the costs and risks of making mistakes. But there have been many casualties. Many retailers closed. And Amazon is the price setter. Publishers must find a way to maintain their profits at Amazon. Or not.

Of course, Amazon has not stopped at books. It is now a retailer of everything. Its delivery drones seem like a crazy idea, but Amazon is so committed to realising them that it already has a website on which they write. 'We're excited about Prime Air — a future delivery system from Amazon designed to safely get packages to customers in 30 minutes or less using small unmanned aerial vehicles, also called drones. Prime Air has great potential to enhance the services we already provide to millions of customers by providing rapid parcel delivery that will also increase the overall safety and efficiency of the transportation system. Putting Prime Air into service will take some time, but we will deploy when we have the regulatory support needed to realise our vision.'

Any company or industry that thinks they are safe gets caught out. The list is long. The media, clothing, white goods, hotels, taxis, Nokia, IBM, real estate, car sales, recruitment, airlines, travel agents, accountants, second-hand goods. The situations they face won't get any easier as Google, Amazon, Alibaba and eBay expand their businesses by the day.

It's not the big disruptions but the little ones that topple industries

For most industries, disruptions are not so easy to see. We face disruption by degrees; death by a thousand cuts where a cut can be the global incident of the day or a new app for a phone. An app that improves the value proposition for a small and specialised segment of the population can be a threat. Then they get their segmentation more defined and launch a global offering. It can be highly competitive for a niche product or that market segment.

Despite this, there are still industries and companies that speak and act as if they are impervious to digital disruption. They are not. I recently read this headline: 'Peer-to-peer lenders will never challenge the banks …' It reads as if this is the conclusion to the risks banks face.

Banks are fully alert to the danger of digital disruption, according to the Economist Intelligence Unit's report, The Disruption of Banking. Alert, but not alarmed, unfortunately. The authors write: 'While apparently concerned, banks do not appear to be stepping up to the challenge. A majority of bankers (54%) believe that banks are either ignoring the challenge or that they "talk about disruption, but are not making changes." An even larger percentage of fintech (financial technology) executives (59%) agree with them. What is holding the banks back? By their admission, banks see the main barriers to responding to fintech as the 'soft issues'—lack of a clear digital strategy, cultures unsuited to rapid change and an inability to attract top technological talent. 'It is a challenge we face as banks to sustain the entrepreneurial spirit,' says Hector Lagos Donde, president and managing director of Mexico's Grupo Monex.'

SOME COMPANIES ARE GETTING IT RIGHT

These companies are opening themselves to the sheer enormity of digital disruption in bite-sized pieces. I was fascinated to read about a two-year pilot run by the consulting arm of the global accounting firm, Deloitte, because it illustrates a potential new model of digital intelligence.

Deloitte's initiative is a form of crowd-sourced consulting, according to the report by the *Australian Financial Review* journalist, Agnes King. In a radical departure from the traditional strategy consulting model, Deloitte's approach involves breaking down large problems into smaller bits and outsourcing them to 'ecosystem of leading crowd sourcing vendors – including Topcoder, 10EQS, Wikistrat and InCrowd' – and, according to Deloitte's publicity, '… inviting individuals to contribute in a time-bound and skill-specific way, has resulted in faster, better, and often less-costly outcomes.'

As a model, it's an initiative that seems to me full of potential. Here are the key elements of such a model that appear to illustrate the kind of digital intelligence we need.

- Revenue is shared with other expert consultants – the game is no longer dominating the market, but playing in it. A dominant player in the market recognises that disruption is inevitable, and takes some control over that by giving away parts of its business to smaller players.

- Clients benefit from faster and cheaper and better results.

- The dominant player has an opportunity to see where people are going, and what they're paying. It is giving them intelligence and relationships with potential disruptors.

- It demonstrates a confidence that a lot of mature, traditional companies would not have, by saying: 'Well, let's go out there and welcome this new world and work with it.'

- The model lets society determine how it's used – it is open-ended and experimental. In Deloitte's case, they gave it a two-year trial before launching it globally. But the reality of this kind of model is that no-one knows where it will take us.

- It's low cost. It is forfeiting some of today's revenue to preserve tomorrow's revenue by opening the doors to the world of disruption.

There are lots of companies that have survived disruptions to their industry. The American bookstore, Barnes & Noble, after a series of mergers and near-collapses, has rebuilt its book retailing business. Dymocks is another survivor in the post-Amazon era.

But how different the outcome might have been. Had those companies sought out Bezos in the early days, built a relationship with his business when it was a fledgeling, and discovered the enormity of his plans while they still had something with which to tempt the man to open his doors to them. With a clear digital strategy, a culture of openness to personal professional disruption and challenge, could the corporate world today build relationships with the energetic newcomers, the entrepreneurs and innovators and see themselves part of the future? This is a model worthy of exploration. I can provide some comfort to directors by debunking a couple of myths.

BOARDS SHOULD DISCUSS BUSINESS INTELLIGENCE, NOT BITS AND BYTES

Digital intelligence is not a matter of understanding every aspect of modern computer and networking technology. It's not about electrical engineering, or bits and bytes. In fact the idea that directors must understand bits and bytes is one myth that contributes to the problem. Directors know that technology is sophisticated, and they know it isn't possible for them to grasp the ins and outs.

The conversation for directors is more constructive if it is about business intelligence and focused on the information needed within a particular time frame, and what are the critical information assets of the organisation.

That is familiar territory for boards. Businesses deal with the information that supports decisions and business intelligence. What is not familiar is that the online world has created information resources and online environments on a scale that we have never seen before. What we know and understand and make decisions on can now change in a second and can be of a global, national or business significance.

The rate and impact of our exposure to so much global information is challenging. We have to make sense of this 'all-to-all' cacophony, the microsecond intelligence updates, so to speak. That is what we must work with.

TECHNOLOGY IS NOT THE WHOLE STORY

The global phenomenon that is the ride-sharing business, Uber, is not a story of new technology; Uber's success is the story of applying an existing technology to a new market. It's about understanding the behaviours that social networking technology unlocks. The people who captured the Uber opportunity knew the way people work with the new capabilities offered by smart phone technology and social networking. Uber came along, and people said, 'Well the Uber cars are clean, and I know when they're coming. If I don't like the car, I vote against them, then other people vote against them, so they lose their work anyway. If I'm a bad passenger, then the driver votes against me, so I can't be bad.' You've got this self-monitoring, social monitoring thing that is doing a better job than the regulator. How many people did we pay to regulate the taxi industry? Apparently they didn't deliver any more value than we could all deliver ourselves and collectively.

That is a study in social needs. It is not a technology that we need to understand here but changing demographics, changing social values, and how a digital technology unleashed existing and emerging forces and requirements and abetted and enabled some we didn't even know we appreciated. The discussion can't just be one of technology, it must be about the way business relevant information is created and delivered as services via technology.

Given directors responsibility in strategy, and that they can't just leave it to the experts, we need some tips to bring directors back into the action of strategy. What are the keys to getting it right, building digital intelligence on our boards and building our nation into a global technology leader, not just in ideas, but in the whole commercial ecosystem? Fortunately, it is not rocket science.

1. Build digital intelligence and engage in challenging digital strategy

All directors must be confident and informed sufficiently to engage in challenging the status quo and embracing new ideas and opportunities – just one so-called expert is never sufficient for a rigorous conversation. The board's processes of examining the risk and opportunities of a technology-based strategy are no different than for any other plan, or strategic project. What is different is that we need to thoroughly test the advice we are receiving. Vendor interests in what they want to sell to the company, help from the people who describe technology as PCs, servers, screens and networks, and removal of the 'information' part of IT, leaves boards with thin information upon which to make decisions. It is not realistic for directors to test the assumptions such as the costs and time involved in the technology implementation or even if it is current and competitive unless they have the understanding of how the technology is serving the company's information and services' needs. So the board needs to stand up for itself, and ask for the business information behind the technology proposal. It can also help if the board brings external experts into the room for the strategy discussions.

We do need to put ourselves on a learning curve. How can we do that? The Deloitte example referred to above is a good one where they are learning alongside the digital disrupters and participating in digital disruption. Participation in some way is essential. We cannot all be inventors and entrepreneurs, but we can be investors, mentors and sit on the boards of start-up companies. As an angel investor in start-ups, I see the disruption from all quarters, in all industries and social and commercial sectors.

All board members must have a go at understanding the direction technologies, digital and others are taking and what they enable, and not just superficially. Bring in external tech experts to talk about emerging trends and issues. Use it as an opportunity to learn together as a team, and become more confident in talking about potential issues, risks and opportunities and the case studies, good and bad of how other organisations have fared.

We cannot allow or accept that digital intelligence is optional. It is every director's duty to understand what is happening in digital technologies and how they are being applied innovatively in the digital consumer world. Many of us have children only too willing to give us a rapid immersion in the digital world. When listening, playing, experimenting, and professional training and development to build our digital intelligence is mandatory for the director's role, we will see the transformation we need.

2. Embrace a new definition of digital disruption

Digital disruption is about seeing the way societies and individuals are interacting with new devices and capabilities and bringing those together in innovative ways.

Understanding Facebook or other social media is useful but not sufficient in our understanding of technology. There are always discoveries that threaten to be just buzz words – and possibly dangerous in the hands of the partly informed director or to become a game changer for your organisation.

What can board directors do? The first thing is to understand some of the social and conceptual trends that are sitting inside digital businesses such as Uber or Airbnb. McKinsey & Company's recent article on "Marketing's Holy Grail: Digital personalization at scale" is one interesting concept. Gartner, a leading information technology research and advisory company annually dientify the top 10 strategic technology trends. For 2017, these are:

1. Artificial Intelligence (AI) and advanced Machine Learning (ML)
2. Intelligent apps
3. Intelligent things
4. Immersive technologies, such as virtual reality (VR) and augmented reality (AR)
5. Digital twin

6. Blockchain and distributed ledgers
7. Conversational systems
8. Mesh app and service architecture (MASA)
9. Digital technology platforms
10. Adaptive security architecture

Every one of the above presents an opportunity and a risk. To read more about each trend, visit: http://www.gartner.com/newsroom/id/3482617. Board directors don't necessarily have the means by which they can understand all of the technology terms, but they can certainly understand the concepts and the impact each could have on their business, as both an opportunity and a threat, and an impetus to a new game strategy.

3. Information is the new asset

Twenty-five years ago, when I was at the Commission for the Future, we talked about the dematerialisation of society in response to environmental and production costs pressures and looking forward at that time to 'paperless offices' and 'virtual work.' It didn't happen overnight, but it is going on. I don't think we have that long timeframe for this popular revolution.

Organisations have previously retained and invested in what we thought of as the organisation's strengths, such as a physical asset or the assets within the business's processes and systems. Even this legacy thinking may, in fact, be a liability. What does it take to 'de-asset' legacy assets and legacy thinking? How do we know the value of our information assets?

Today, we need to question the value of our traditional assets. Our digital knowledge needs to encompass the digital information used and created by the company and what has value and contributes intelligence. We need to be supported by the CIO whose role has shifted from being operational to being the gatherer of strategic digital intelligence.

4. The interaction of people, society and technology is a critical influence on the future

How can board directors guide their organisations? One can look at business trends, although unfortunately, it is historical and not always relevant by the time we see them. One can look at people trends – at customer behaviours and factors that are influencing their buying decisions. One can look at the technology trends, such as the Gartner trends discussed above, and challenge your organisations to think through what impact – negative and positive, there could be, taking one technology trend at a time. However, society is surprising and unpredictable, and people's behaviours are influenced by and in turn influence what technologies are adopted and for what reason. Scenario planning is the only way the confluence of social, technological, economic, environmental and political forces can be investigated and understood.

5. Boards must engage in revealing the 'new game strategy'

In these disruptive times, the bigger risk is not whether a board can properly test and challenge an information technology based strategy brought forward by management. It is whether boards can challenge and guide the management team to see over the horizon at the future challenges, opportunities and new ways of conducting business and serving customers and make sure that the organisation is considering how technology and the changing society will open the way for a new strategy. Seeing into a void is always harder than examining the evidence in front of you.

So what does it take to derive a new game strategy?

- Courage because looking into the future and taking a step there is frightening
- No sentimentality for traditional approaches that have served you well to date
- Knowing and challenging the assumptions that sit behind your understanding of your business model, customers and products or services.
- Strong understanding of what is happening in technology and new business models, i.e., where the disruption is coming from and how fast

- And linked to this, a firm understanding of where people and our societies are changing as they react with the new business models and technologies. So we don't necessarily all need to be IT experts to be able to participate knowledgeably in formulating strategy, but we do need to be digitally literate, digitally curious and prepared to analyse and understand what is happening at the leading edge

6. One tech expert won't do the job

A big mistake boards often make thinking a single expert – or woman, or digital specialist – is enough to effect change. It takes more than one person to have a conversation about digital disruption. Boards cannot debate digital disruption in their industry with one person sitting at the boardroom table bleating about disruption and getting shouted down, or, more often, politely ignored. To foster a conversation, everyone needs to have the confidence to hold a view.

7. Proactive external engagement and inquiry

We must be willing to ask questions until we understand the answers. It's a matter of learning to ask the right questions. What is leading edge thinking and research in our start-ups, our research labs and universities? What is attracting the attention of consumers and why? How might that trend impact our industry or company? What flaw in our business model do we take for granted that a disruptive company might use against us? How can we build relationships with new young companies?

Talk to people. Go to universities, to the CSIRO and other technology specialists, and consulting companies. More importantly go to the incubators and accelerators. Find the supply chain for challenging new ideas in your industry. Ask them what's going on, what's new, where is the cutting edge on this and in this sector? Who is using this sort of technology? How are they using it and what is its potential relating to us?

8. Let's not accept the language barrier

We need to break down the language barriers between the IT experts and business experts. We have had some clumsy ways of doing that in the past. As a result, we have dumbed down our technology approaches. We have run a lot of risk by taking the experts out of the decision making by preferring the chief information officer with business language skills over the CIO with deeper skills and technical abilities but fewer business communication skills

Over time, we've developed an 'IT or technology' language. Jargon is what stands between what we say and other's understanding of us, the author Teryl Burt explains in his book, 'I Don't Speak Geek: a simple guide to help businesses navigate today's complex technology choices.'

'Business speak' is much the same – a jargon-filled shorthand that makes sense to business folk but no-one else. So these two separate languages are not understood between the parties.

We need to put an end to that. As directors, we have to be willing just to say, 'Okay, I don't understand that. Please explain that to me again.' That is the way to acquire real digital intelligence. It is directors being proactive and making sure we're all talking the same language, and we are moving towards the same destination instead of allowing that divide to continue to happen.

9. Bad experiences, bad habits

We have formed bad habits in the past. People have made a lot of mistakes in IT projects that have been very expensive. As a result, they've said, 'Okay, let's just buy it off the shelf. We don't want to do anything new. We don't want to invent anything. We want to buy it off the shelf because we think that that minimises our risks.' But if you are buying it off the shelf, with the technology provided, then you haven't done anything new for a long, long time.

We have built bad habits out of bad experiences, and a misguided approach to risk is the result. But the real danger is not about not making expensive mistakes. In fact, we are running the risk of not being there with the latest and the newest and the creative edge to the thinking. That's a bigger risk.

10. Diversity

We must fill our boardrooms with as many perspectives as possible. Diversity in education, experience, gender, age and culture are shown again and again to deliver more ideas and bigger profits.

11. Determination

Failure is not a comfortable experience, but it cannot be avoided completely if we are determined to build our digital intelligence. We must be determined to take calculated risks with everything from our products and services to our business models, to our approach to innovation.

12. Attitude: Let us be the first

Being a technology follower is a strategy that served many companies well in the dotcom crash. We learned a lesson: early adoption is risky. But that is over 15 years ago. That is not the strategy for our times.
We must be willing to be first to market.

I have no doubt that we are capable of overturning current uncertainties, building our confidence and competence at the board level, and becoming digitally intelligent.

We have good examples to follow, and to it's not that hard to achieve. In fact, it's exciting and opens up a world of opportunity. It's about a shift in attitude and a change in skills. The first step is to commit. The second is to recruit cleverly. And the third is to change the company attitude from 'let's follow' to 'let's be first'.

13. Think like a technology company

Every organisation is arguably a technology company. If the board is not considering how technology can transform the organisation or how it can disrupt it – it's not thinking about its future. But caution! Traditional thinking has said that IT is about automating business processes like HR, billing, manufacturing, office automation and databases. This is an industrial age paradigm. When a company is an IT company, it is their business language, methods and working environments.

If we approach business by thinking like a technology company, we will not relegate the IT and information engineers to implement strategy but will invite them to lead it. What are the digital and information assets that create the power of our business and how does that digital power evolve our business on a global scale into the future we want?

THE FUTURE IF WE DO NOT ADDRESS DIGITAL INTELLIGENCE

How many of us know that an Australian invented Wi-Fi. John O'Sullivan, an Australian electrical engineer, invented the core technology that makes a wireless local area network fast and reliable. The nation's government-funded science organisation, the CSIRO, patented the technology that forms part of the Wi-Fi standards.

Australia did not commercialise Wi-Fi. In fact, the CSIRO has been engaged in a long legal war to protect its patent and get our nation paid what it is due. That is good, but it is not sufficient. We need to spend less time in law courts and more time in skunkworks. We need to be part of the story of technology. Instead, Australia was left out of the story of Wi-Fi. An article by The Economist, A Brief History of Wi-Fi, mentions neither Australia nor Sullivan.

We cannot commercialise our technology if we have neither the digital intelligence, the technology 'eco-system' (think Silicon Valley), the technology investment or the confidence and curiosity to reach into the community of innovators and entrepreneurs, scientists and researchers, to support it.

THE FUTURE IF WE GET IT RIGHT

Our manufacturing industry, in decline for decades, is facing a more positive transformation. The invention of 3-D printing – machines that can print objects rather than text or picture – will return manufacturing to our shores. We can already print emergency housing and customised push bikes. Last year, a cancer survivor received 3D printed ribs, made of titanium, for the first time. The future is leading us to unimaginable places.

I recall Jim Dator, a well-known Hawaiian futurist and recreational surfer urging us to ride the tsunamis of change all the way to the shore. That could be a terrifying call to action or a lot of fun

Susan Oliver is an experienced chairman and company director. She is the founding chair of Scale Investors, which invests in women-led start-ups. She is an independent member of the Investment Committee for IFM Investors, a leading investor in global infrastructure for industry super funds. An entrepreneur, she is the co-inventor of two software systems.

Susan has served as chair and director of listed companies since 1996, including Transurban Group, the restructure of Centro, MBF (now BUPA), Just Group, Programmed Group and Coffey International.

Susan's extensive experience helped shape national policy agendas in innovation, information technology and arts. In 2001, the Australian Government awarded Susan a Centenary Medal.

She is a passionate supporter of several not-for-profit causes. She serves as chair of the Wheeler Centre, and is a director of the Melbourne Theatre Company and Melbourne Chamber Orchestra.

She helped found The Big Issue in Australia and sat on the board of The Smith Family.

CHAPTER SIX

THE SPIN IN

WHY OUR INNOVATION MODELS
FAIL US AND HOW TO CHANGE THEM

BY JOE WARD

Founder, Acceleration Bay

ABOUT 15 TO 20 YEARS AGO, GLOBAL companies started an unusual strategy: building patent portfolios. Until then, chief executives of global companies focused on growing revenue or cutting costs or both. Intellectual property was the responsibility of the in-house lawyers. It was a necessary expense. Building intellectual property banks was the domain of venture capital companies. Today, the situation is very different. Global companies have built enormous patent banks as part of their strategy to defend today's revenue and innovate to build tomorrow's sales.

The three biggest patent holders in the world – Samsung, Canon and IBM – own a combined stash of 142,881 active US patents. The strategic reasons behind the patent banks are strong, and I will explain them later in this chapter. However, problems are emerging with patent banks. They are expensive to build and maintain. They are even more expensive to defend. And they are full of wasted potential. That potential goes beyond financial opportunity. And it goes beyond the opportunity to develop these latent ideas into products and services that our world desperately needs.

Global companies face two problems in realising the full value of their intellectual property. The first is that big companies struggle to create successful 'spin-off' companies based on their intellectual property. A spin-off company is one that is started within the parent company, and sold to recoup the investment. The model often fails because the structure and overheads of the parent company are incompatible with the agility the spin-off needs for success.

The corporate Goliaths face a second problem that might come as a surprise to some: they struggle to maintain control of their intellectual property. Much is written about big companies ripping off intellectual property from smaller, weaker rivals, who cannot defend themselves against the deep pockets of the global corporation. But the reverse is also true: start-ups exploit public sympathies to plunder and use other company's intellectual property without paying license fees. Big businesses have the funds to defend infringements, but most choose to avoid the damage to their public image resulting from chasing small companies for payment.

The 'Spin In' is the answer. The Spin In offers a new commercial model for big business to unleash the potential of its intellectual property, and create a positive collaborative model with start-ups. In this chapter, I'll tell you how it works, and why it's a great idea: for global companies, for ambitious entrepreneurs, for investors and for a planet that is in desperate need of technology solutions to its many problems.

The ideas in this chapters are born of experience. I'm a serial entrepreneur and have spent decades at the frontline of innovation. I've seen success and failure. I was among the first two employees at one of Australia's best-known internet service provider start-ups, Ozemail (backed by Malcolm Turnbull during his career as an investment banker). I founded a pioneering company in online procurement, MarketBoomer. I've seen everything that can go wrong. I've kept many friendships despite the enormous pressures of building new companies, but I've watch others suffer the personal cost of failure.

I see a colossal entrepreneurial opportunity that is locked up inside big companies.

Neither they, nor the world, are benefiting from this opportunity. I want to see that change. Now is the time to reinvent the innovation models that power our biggest companies. The models we are using are broken. They do not serve the entrepreneur, the corporate or the world –

at least not to their full potential. These patent divisions are led by some of the smartest minds in the world of intellectual property management and development.

This chapter is for everyone who wants to unleash the potential of all the minds that have combined to build this ideas bank. In it, I explore a new innovation model that uses creativity and collaboration to build value, protect IP, make money and serve humanity.

HOW TO CLEVERLY TO UNLOCK IDEAS HELD BY BIG COMPANIES

In the year 2000, the then chief executive of Xerox Corporation, Richard Thoman, made headlines in the *Harvard Business Review* by announcing his intention to focus on intellectual property. The *HBR* authors reported:

> Richard Thoman is not your typical chief executive officer. Most Fortune 500 CEOs, when asked how they intend to increase shareholder value, will talk about increasing sales, creating new leading-edge product lines, or pursuing mergers and acquisitions. But Thoman, who was appointed CEO of the $20 billion Xerox Corporation last summer, isn't content with such conventional strategies. He believes one of the strategic keys to Xerox's future is something so intangible, so invisible to traditional bottom-line thinking and corporate practice, that it doesn't even show up on the balance sheet.

Thoman's idea wasn't new, but it was rare at the time. Global computing company, IBM, where Thoman was chief financial officer, had an aggressive intellectual-property program that, *HBR* reported: 'boosted annual patent-licensing royalties a phenomenal 3,300%—from $30 million in 1990 to nearly $1 billion.'

Over the ensuing decade and a half, other corporates followed suit, building vast banks of intellectual property. Their strategies for using these assets straddled several critical goals, outlined in the *HBR* article, which I summarise below. These companies used IP to:

1. Establish their market advantage by:
- Protecting core technologies and business methods
- Anticipating shifts in the market and new technologies
- Product innovations that maintain their brand domination

2. Improve their financial performance by:
- Tapping patents for new revenue
- Reducing the costs of maintaining patents with better management

3. Enhance competitiveness by:
- Outflanking competitors
- Exploiting new market opportunities
- Reducing risk

Companies fail to gain an innovative edge

Sadly, these once-clever plans have lost their strategic edge. How do I know? Research and development spending across the world is grinding to a halt. The total expenditure in 2016 is still an impressive amount – $US 680 billion, according to the 2016 Global Innovation 1000 Study by accounting firm, PwC, which publishes an annual survey of trends among the world's 1000 largest corporate research and development spenders.

Over the past five years, the Chinese technology company Huawei has spent nearly $US30 billion ($40 billion) on research and development, The Australian newspaper reports. In 2016 alone, Huawei's R&D spending rose 46 per cent to $US9.2bn, beating the $US8.1 billion Apple spent in its most recent fiscal year.

The glacial growth of research and development spending

But the percentages tell another story. The total amount, however impressive, is a miniscule increase in the total R&D spend: just 0.04%. What can we learn from this figure? The old innovation model is dying. More startling is another fact reported by PwC: there is no longer a correlation between innovation and the amount spent on R&D. Let me say again: *no correlation*. As PwC reports.

> *The 10 Most Innovative Companies continue to outperform the Top 10 R&D Spenders on key performance metrics, as has been the case for each of the past seven years.*

Building a patent portfolio is expensive. Between 1984 and 2014, the once-great phone company, Nokia, invested more than €50 billion to create a portfolio of 30,000 patents and patent applications, according to the online publication, *Techspot*. (In the end, these patents became the most valuable assets in the company).

Patent wars intensify

But despite their value, keeping hold of patents is a cutthroat business. Massive laws suits, patent trolls, and litigation in the communications technology sector reached a record high over the past decade.

In 2016, companies filed nearly 3,000 technology litigations, according to the legal website, Lexology. High-profile legal skirmishes damage the public perception of the parties involved, even when both are Goliaths.

And defending patents is also costly, tech journalist, Jim Kerstetter reports. The median legal cost for a claim that could be worth less than a $1 million are $650,000. When $1 million to $25 million is considered 'at risk', total litigation costs can hit $2.5 million. For a claim over $25 million, median legal costs are $5 million or more and continuing to rise with Inter Parte Review challenges.

The rise of the patent trolls

The damaging PR arising from patent battles led to the rise of a new breed of company – the 'patent troll'. Patent trolls have a terrible reputation in the innovation community as blocking, not fostering, innovation. The patent trolls' make money by buying patents and enforcing infringements. They are like the companies that speculate in buying and selling domain names.

The patent trolls are universally hated by start-ups in Silicon

Valley, where I live. But we should not be surprised by their rise. They are another symptom of a system in crisis or, more truthfully, in its death throws. What are we to do about this demeaning descent into an unhealthy, costly legal merry-go-round that does little but ensure that many great ideas do not see the light of day?

HOW TO FOSTER FUTURE INNOVATION

Thanks to the patent trolls, investment in IP is a controversial idea. But the spin-in model is distinctly different from trolling. It is not about legal skirmishes, but about finding entrepreneurs who can build real companies from intellectual property, and defend the patents on real commercial grounds.

To understand this new model of innovation, we need to be clear about the current model – the 'spin out' – which is the mainline of innovation by global companies. Let's unpack both models.

The Spin Out: Stage One

The spin out is a two-stage innovation model. To get started, corporate leadership selects some of its intellectual property and invests in the early stages of development. It sounds simple, but it's harder than it seems. A good idea might be buried among thousands of employees. IP lawyers are tied up in court. Anyway, most lawyers wouldn't recognize a commercial idea if they fell over one (and some of my best friends are lawyers). With so many ideas in their banks, it's expensive to test all of them in the hope of finding a good one – the spray and pray method of IP development. The alternative – choosing select ideas and conducting feasibility studies is fraught.

To be brutally honest, most corporations are not known for their entrepreneurial nous. I'm sure most would admit it publicly. The talented individuals who elect a career in our world's biggest companies have different skills, motivations and experience than those people who excel at the lean start-up.

Typically, spin-outs are the result of 'accidents' – a fortunate meeting amongst thousands of staff, a Steve Jobs lurking in the technology transfer division, the chief financial officer reading the business pages about emerging trends and suggesting the IP lawyers scour the banks for opportunities.

The Spin Out: Stage Two

Once found, the clever IP is commercialised in-house in a special division, often less encumbered by lumbering bureaucracy of the rest of the company. Millions are misspent, but if the product makes it to market, the corporate is faced with an immediate problem: can it afford to wait until the in-house division is profitable, or should it sell to recoup its millions.

When the decision to sell is made, the baby company is 'spun out' of its parent and sold, most often to venture-capital investors and the corporate executives who have developed it far enough to believe in its future.

The risks and rewards of the spin out

Risk potential
The new company will fail. If the business succeeds, the parent company loses most of the value of its investment when it sells the company. The future growth and the biggest profits are sacrificed for immediate return on investment.

Return potential
The initial investment is recouped when the company is sold, hopefully with a substantial profit. Let's turn again to HBR for an example of a good spin out.

> *The aerospace firm Lockheed Martin ... over the years had assembled a large cache of 3-D flight simulator patents that gathered dust in the corporate legal office. But in 1997, the company used those patents as the foundation for a new venture called Real3D that it spun off to compete in the PC graphics and video game business. Real3D attracted investments from Intel*

and Silicon Graphics, and it's currently valued at several hundred million dollars. Lockheed took a group of fallow patents valued on its books at exactly zero and transform them into a strategic presence in a potentially lucrative new market. It also gained a 40% stake in a high-flying start-up.

THE SPIN IN

The Spin In: Stage One

The Spin In is a three-stage innovation model. The company sells its intellectual property to a corporation like Acceleration Bay, that specialises in the innovative strategy of the spin-in. Acceleration Bay and others are experts in sifting through patents for untapped innovation opportunities. We know this because many universities open their patent banks for this reason. Universities often demand royalties in return for their intellectual property – a problematic model, since royalties accrue from sales, not profits. Once the companies like us finds an opportunity, they buy the patent and the rights to develop it.

The Spin In: Stage Two

Companies, such as Acceleration Bay, match the idea with an experienced entrepreneur with a track record of building new companies in the right industry or sector. Assembling a small team, possibly one they have worked with on previous spin in ventures, the leader starts and builds the new business using methodologies outlined by Eric Reis in his bestselling book: *The Lean Startup*, and making the most of the networks, contacts and money from its backers.

If other companies infringe the patent, the startup will defend it using funds from its backers, and the expensive process of enforcing the patents and protecting the value of the invention commences.

The Spin In: Stage Three

At a point agree to from the start, the parent company can exercise an option to buy back the startup. The milestone that triggers the option might be revenue, profit, or number of staff – whatever the investors and the parent company agreed to from the start. The parent company uses its strengths – people, resources, clients and brand – to grow the startup, either keeping it as a division, or building it until it can be listed.

The risks and rewards of the spin in

Risk potential
This model drastically reduces risk for the parent company that banks the IP. The parent company receives an upfront payment for the patent portfolio and faces none of the risk of developing or licensing the assets. Conversely, they face no risk from the idea being commercialized – and possibly competing with the parent – because they have the first option to buy back the startup.

The spin-in investment company takes the risk on development, but they are experts in this kind of risk. Their speciality is to value the risk and balance it against the potential reward. Their risk is lowered by the prospect of a buyer to recoup their investment once the option is triggered.

Return potential
If the startup is successful, the investors make a return on sale back to the parent – the spin in. The parent makes money when if first sells the IP, sidesteps the riskiest stage of development, and only pays for the business once its feasibility is proved. It can recoup this second, more substantial investment by adding the new company's products and services to its sales revenue, or by building the company ready for listing or resale. I can't provide the details here, because they are commercial in confidence, but at the time of publication, Acceleration Bay had secured $US 120 million in transactions.

Intellectual property lawyers have more to offer the world than being the legal guardians of patents. The spin in is a creative opportunity for IP professionals, a chance to work on exciting projects that provide solutions to the world's problems and vastly increase the value of their patent banks.

CONCLUSION

There are billions of dollars held in patent portfolios around the globe. Right now, they are stuck because the research and development model is broken. Long term investment in research and development is falling, which is worrying because companies must innovate to survive, and the world needs new ideas urgently; we have a lot of problems to solve.

The old spin out model of starting new companies based in intellectual property is losing popularity because it doesn't make financial or strategic sense – corporate structures are not compatible with the agility that startups need to succeed.

That leaves intellectual property lawyers in a one-dimensional roles as the guards of the patent banks. Most of the money to be made is from infringements, and prosecuting is negative PR for big companies

The Spin In offers a new model of innovation that radically transforms the relationship between global companies and startups. They become partners, not adversaries. It shifts the innovation culture away from the current defensive litigation focus towards a value creation model that unleashes new ideas onto the market and rebuilds the business case for research and development.

Little stands between us and this new model. Perhaps the biggest risk is that companies, large and small, stay captive to redundant ways of operating simply because of culture – 'that is the way we do things around here'.

That is not good enough. The most talented professionals gravitate to companies that innovate. Customers search for companies that can prove a link between social good and financial gain. Whatever the barriers, it is time to tear them down and move towards this new innovation model. Now.

CONTACT ME

If you want to find out more about this new model of innovation, email me at Acceleration Bay and put 'Future Makers' in the subject title. I'd be delighted to hear from you. Or send me a Tweet.

joe@accerationbay.com
@accelbay

Joe Ward is a serial entrepreneur with over 30 years in media and technology ventures. A network engineer by profession, Joe is now a thought leader in business to business technology ventures.

Based in San Francisco, California, Joe is the founder and president of Acceleration Bay, a ventures corporation that creates patent-backed ventures. His platform creates ventures based on inventions by large multinational corporations in the networking technology vertical.

Joe's career highlights include the build out of the largest computer publishing network in the Southern Hemisphere for Kerry Packer's Australian Consolidated Press. Later moving on to leading technology ventures in B2B Procurement and eLearning.

After relocating to Silicon Valley in 2009, his new venture successfully engaged with News Corporation in New York, deploying his venture's networking technology into Fox News and Myspace.

In 2014, he founded Acceleration Bay, which partners with companies such as Boeing to continue his passion for B2B technology ventures in the mesh networking sector.

CHAPTER SEVEN

DIGITAL TECHNOLOGY: IT'S AN ACTION SPORT

BY PETER WILLIAMS

CEO, Deloitte Centre for the Edge

HOW DO YOU BECOME A FUTURE MAKER? In my case it was a mix of restlessness, discovery and curiosity. I moved to the United Kingdom in 1993 as part of a work secondment. I had been working as an Insolvency accountant at one of the Big Four accounting firms, Deloitte, for 10 years and, while I was good at it, I felt I wasn't growing professionally speaking. At that time, we noticed the emergence of the World Wide Web and the launch of one of the first the browsers: Mosaic. I was reading about the 'Internet' and 'Information Superhighways' and had no idea what it all meant. In London with an hour to kill before a meeting, I saw a café with a sign 'Internet Here'. I went in and asked the proprietor to show me the Internet. After a few minutes of looking at various websites, I turned to the guy and said, 'This is going to change everything!' My curiosity instantly turned to a passion and I immersed myself into everything web and internet. My orientation was partly 'How good is this?' but also 'How far can this go?'.

Whenever I look at new technology, I immediately start to think about where this technology is heading and what it will mean.

FROM INTERNET CAFÉ TO INTERNATIONAL NETWORK

When I got back to Australia, I convinced the powers that be at Deloitte that we needed to set up an eBusiness consulting group. I transferred to a consulting unit and that group eventually became Deloitte Digital, which now operates in more than 30 countries around the world. When I reflect on its humble start, in an accounting firm in Melbourne, it was really a combination of passion, belief and commitment … and a lot of hard work. In addition, it was reaching out and connecting with people who were similarly passionate and wanted to make an impact. The other key ingredient was a desire to learn, and to find ways we could apply what we were learning and thinking, that set us on a pathway to a sustainable future.

Digital is an open playing field, and everyone can participate. It constantly evolves and it's not set in stone. There are a lot of unexplored territories that can be combined with new technologies and business models. Historically, technological advantage was about capacity to outspend your rivals to get access to new technologies. These days with easy access to technologies such as cloud, open-source software, smartphones, and machine learning (the list goes on), it is much more about how one company can out think and out execute another. Rather than spending hundreds of millions of dollars on technology transformation, the smartest companies can prototype and test new ideas and scale those that show the most promise.

When I started, I thought the most important thing was to learn to code. I did that at the start, but quickly realised that I would never be great at it. What I learnt over the years was that success came from a diversity of skills. Success is about combining people with ideas, ethnographers, artists, designers, project managers, testers and coders. It's a broad field, and everyone can play, but the most important thing is to learn by doing. It's an action-sport. It's not an intellectual, Excel spreadsheet, PowerPoint, or business case project. It's about doing stuff.

THIS CHAPTER IS FOR THE DOERS

If you are reading this chapter, this is my message: I would love to see you using what is written here to do stuff. At the Deloitte Centre for the Edge (the advisory unit of Deloitte Digital), we write a lot of 'thoughtware' – reports on new technologies and trends – but sometimes I wonder whether our reports are just being used as paperweights or stored as a digital file that 'I must read someday'. So, it is gratifying when I see people referencing the work we have done or hear how some of our thought have been applied. What I'd like to hear people saying after reading this chapter is, 'Hey, we used this information, and it helped us achieve x, y, or z.'

Every day, I meet passionate people whose job is to convince people in high places that digital is important and we need to embrace it. If you are one of those, I'd love you to look at this chapter as a third-party reference. You can say to the next sceptic you have to deal with, 'Here! Read this.' What I am talking about in this chapter is written for the people who, like me, over many years have been trying to get people to see what's going on. It's for those people whose bosses says to them, 'That's all good, but prove it.' These days, it is much easier to persuade our bosses to listen because we are surrounded by technology and people using it in so many different ways; it is difficult for anyone to deny that something is happening. Back in the mid 1990s, it was a lot tougher but persistence combined with getting stuff done in spite of the barriers you might face is how you can win the day.

When you're a consultant for a long time, as I have been, you find you are often telling your client nothing different from what the people in that organisation are telling them; it's just that the consultant is saying it, not somebody inside. I learned to take the attitude that it doesn't matter who gets listened to; it is much more important that we get things moving. Knowing that means sometimes you need to leverage external people or respected journals, even a well-delivered Ted talk, to get things moving. Another way is to show that a competitor has picked up your idea and has run with it and is getting a competitive edge. Rather than feeling like you are belting your head against the wall, understand that innovation isn't just about ideas and execution, often it is about managing politics and winning people over and you have to be innovative in how you manage that as well.

I've worked with many leaders over the years. Occasionally you find a visionary leader who knows broadly what they need to achieve but is not clear about what to do and how to do it. They set big goals and empower teams to get on and do it. These people are rare gems that bring the best out of people and change the art of the possible. I worked with a CEO at an insurance company who realised he had to get his organisation moving on digital. He said to the team, which included a combination of his internal leaders as well as a number of external companies that were about to start working on the project, 'We need to get serious about digital or we won't have a business. We need to learn to move fast and break the way we have traditionally done things. We also need to take what we learn and spread it through the organisation'. It was the only project I had done in more than 20 years of digital projects where the output was not only a world-class customer and employee experience but also a collation of what we learned on the way, which could then be adopted by the whole organisation.

I am seeing more people in leadership roles who are getting serious about digital. For many years, they have heard the narrative that 'digital disruption' is a looming threat. It has got to the point where they are saying,

'Yeah, we get it. Digital disruption was supposed to wipe us out but we are still here. What do we do about it? I know I've got to do something, but I don't know what to do and I don't know how to think about it.'

That's the executive I'm after.

For these executives, their biggest problem is the capacity of their organisation to innovate. Historically they have looked at transformation as a long slow process over a number of years with a 'big bang' approach to launching the changes publicly. Chief executives or, more recently, chief transformation officers, regularly come to me and say, 'We're in the midst of a transformation project.' (It's usually X number of years multiplied by $100 million, e.g. 'A five-year, $500-million transformation project.')

'Okay, good. How far in?'

'Two or three years.'

'How long to go?'

'Can't see it ending. Can't see the value. Nothing is happening, but in the meantime our environment is changing fast and we don't know what to do.'

MOVE AWAY FROM 'BIG BANG' THEORY

My colleague John Hagel, who is co-chairman at the Deloitte Centre for the Edge, says we need to move away from the 'big bang' transformation model. He suggests that large organisations should take a leaf out of the start-up playbook and start working towards Minimum Viable Transformation, or MVT, which is a similar but different idea to the Minimum Viable Product (MVP), a development technique in which a new product or website is launched with sufficient features to satisfy early adopters. A MVT implies that we are not just creating new products but looking at changing business models as well. It involves redefining the business model and adopting an iterative rapid prototyping model with a 'commit, test, learn' approach, followed by scaling up what works.

One such transformation initiative we started at Deloitte Australia in the mid noughties was called Sleepworks. It came from the notion that our business model was based on 'make money while we work'. The quest we were on was to work out how we could 'make money while we sleep'. This laid the foundations for how Deloitte Australia has incorporated digital into many of the services we provide to clients in 2017. Many iterations and many learnings got us to where we are today.

I often call the scaling stage 'deliberate innovation'. After we have learnt by doing (the commit, test, learn approach), then we say, 'Let's take that hill'. Like the great military leader, Napoleon, said 'If we set out to take Vienna, take Vienna'. Let's commit to what we're trying to achieve, and apply what we have learnt along the way. To me that is the secret sauce.

Just so you don't get distracted from the main game of the technology sport, let's bust a few myths about digital. These really make my blood boil. I'm going to name these myths so we can get those out of the way and then I am going to tell the truths.

Myth #1: Robots and software will take everyone's job

That's bullshit. The boring, repetitive, tedious things will be automated, unleashing our creativity, understanding, and creating better experiences. Throughout history new technology is accompanied by dystopian visions of machines taking everyone's role. It even has a name: The Luddite Fallacy. It is fuelled by fears that computers will become smarter than us and often based on the misconception the artificial intelligence is the same as human intelligence, which it isn't. Overall technology has a positive net effect on job creation. The myth also doesn't take in to account the many new jobs that are created in areas that we don't even know about yet. My previous role as a CEO of a web/mobile development firm didn't exist when I was a kid, nor did the roles of the thousands of people employed in Deloitte Digital across the globe.

    I lost my first job as an accountant in a steel foundry when I was 19. I could blame it on automation, but the reality was that the management of the company had under invested in technology for more than 50 years, hiding behind tariff walls, that protected them from overseas competitors. When those protections were removed, and international competitors, who had adopted technology, could sell what we made at prices lower than we could make things, we went out of business. Obviously, re-platforming -- such as going from horses to cars, or from analogue to digital – means job losses in the old industries but opens up jobs in the new. To secure your future, commit to lifelong learning. As an accountant who became a digital guy I can attest to the benefits.

Myth #2: Everyone needs to learn to code

The mantra espoused by so many people in and around the technology industry has spread to our politicians. There seems to be this ridiculous belief that if you can't code, you can't participate in the digital world. To be fair to those who are saying that, one of the first things I jumped into when I first went onto the web was to teach myself to code HTML in Notepad. I mistakenly thought that if I couldn't code I couldn't participate.

So, I learned to code and frankly, as it became increasingly complex, I hated it and was never going to be any good at it. I managed to make my way in the digital world by focusing on what I was good at which, in a nutshell, was the ability to apply a digital lens to a problem and come up with a vision of where we needed to go. The next step was to build teams that could deliver, which is why my thinking evolved.

As this mantra spread I asked my top ethnographers, artists, writers and designers whether they could code. Some had exposure to it, some had never done it in their lives. Without their talent, we wouldn't have been able to do the work we do.

I have no problem exposing kids to some coding in the curriculum, but if they are not inspired by it, don't force them to learn to code throughout their entire education. What is important is that we educate our kids in digital literacy, which means understanding how to take advantage of opportunities presented in the digital world. Sure, we need great coders and I have been privileged to work with many world class software engineers who can do amazing things. What is far more important to understand is that digital succeeds when there's diversity. This means diverse teams made up of ethnographers, designers, strategist, marketers, software engineers, a wide range of technology disciplines including coders, architects, network engineers, data scientists, project manager and testers. There's a whole raft of things that need to be brought together and although we have no idea of what sort of new knowledge will emerge, there will be no shortage of opportunities.

Myth #3: Technology costs a fortune

The good news here is that technology has a 'price performance curve'. That means that over time, price tends to collapse, and performance tends to go up exponentially. Look at bandwidth, or storage, or computer processing power. Organisations play it safe, and tend to buy technology that is expensive and has a lower performance capability rather than using something that's open source (a philosophy that promotes the free access and distribution of an end product). People might say, 'Oh we must have a specific operating system.' Or, 'We need to go and buy an expensive software application because we can't trust open source.'

Well, why not? There's a lot of great open-source stuff out there, which means that your cost isn't really in the dollars … Even with the Cloud, it used to cost a lot to get access to computer processing and servers and all that stuff. Now they're in the Cloud and you pay for what you use. It's moved to a consumption model from a capital-expenditure model.

The dynamics have changed. Why don't we leverage that? Why don't we get aggressive around adopting this stuff and trying it? Again, you can 'learn by doing' in relatively innocuous areas where nobody is going to die, and stretch the capability of your budget. What used to cost millions of dollars now can be done for a fraction of that price. You don't need to spend money on the technology but you do need to spend it on getting the experience right. That's more valuable. In the old days, it was hardware and software applications and configuring that drove the cost. These days you get a better return on investment by spending more on the front end (the bit that your end-users see) and understanding those end-users and the experiences that you're trying to create for them, and the contexts in which those experiences occur.

That's where this massive change is. We haven't moved to this change, yet. It's almost like the world operates in a linear model, and technology changes on an exponential model. So there is a big opportunity here.

Myth #4: You need to go to Silicon Valley

If I wanted to retire, I could make a fortune as a Silicon Valley tour guide organiser. I get one call a week, 'Oh, we're taking the board and executive to Silicon Valley.'

'Okay, why?'

'Because we want to go and see Google, and Facebook, and Amazon, and Airbnb.'

'Have you walked around the block?'

'What do you mean?'

'Well, I know where you guys are, four doors up is this mob, and around the corner is this mob. You're going to rush over to Silicon Valley, you're going to get the Disney tour if you're lucky. You'll come back and you'll think, "Oh, wow. What can we do?"'

There's an enormous amount of innovation happening in Australia but people have this idea that it's a Silicon Valley phenomenon and we need to do our pilgrimage. I said to our company secretary the other day, 'It's like if you converted to Islam, would the first thing you do go for a pilgrimage to Mecca, or would you go to the local mosque? You converted to Catholicism, would you go to the Vatican or would you go to a local church?'

They feel as though the Silicon Valley experience is going to be life-changing. They come back and don't understand how to engage locally.

Myth #5: We need a Silicon Valley entrepreneur here

That is the other myth about Silicon Valley – only they can run our digital transformations. But the Silicon Valley entrepreneur comes in and say, 'My God, I've never had to do a 7,000-page business case before and wait 12 months to make a decision.' They rarely last. It's the old cultural cringe that we used to talk about. The answer goes back to engaging locally. Sure, engage globally but also engage locally. You don't have to run off to find out about this stuff because there's world-class stuff happening right here.

Myth #6: This new technology is a fad

The idea that new technology and social media are a waste of time (and the Internet's a fad) is the dismissive attitude that I've lived with historically. The way I look at it is this: if I see technology that I don't understand, I don't just dismiss it and say, 'That's rubbish.' I think, 'I wonder what's going on there? I need to understand a bit more of it.'

Back in '93, when I went onto the Internet for the first time, I started to realise, 'Wow this could really change things'. I didn't say, 'Oh gees, it's slow, and it's clunky, and it's really difficult to get on.'

Technology continues to move. Don't look at some new technology and dismiss it as if that will make it disappear. That is silly. Understand where it's going, the trajectory it's on. Think about different ways to use this. Explore with a curious and open mind. Over the years I have had people say stuff like, 'Don't tell me anybody will ever buy insurance online.' Or 'If there's ever a day people have a video call on their phone I'll eat my hat.' See what is, but then think about how it's moving and where it's going. The immediate dismissal of it is head-in-the-sand stuff. The core part of this attitude is a notion of curiosity and exploration.

Myth #7: Market capitalisation is the metric that matters

The idea that market capitalisation is the metric that determines what technology will change the world is 'over hyping'. There's a lot of over hyping, in the technology world. Going back to the dot com boom, the narrative was that anything that had a dot com name was automatically destined to succeed. That all came crashing down in April 2000. For the next few years the narrative was much more 'See we told you the internet really is just a fad'. Both were wrong. A current example of over hyping is blockchain. I am hearing the same stuff I heard in the late 1990s: 'Blockchain is the answer to every problem' and 'Look at the market capitalisation of this or that coin' and 'There has been so much venture capital invested it must be all true'. It's the opposite side of the coin to the dismissiveness discussed in Myth #6.

Sometimes people are dismissive; other times a technology gets so over hyped that you have to step back and apply good judgement. The approach is the same whatever attitude you confront: go out and explore, and learn, and understand. Say, 'Well, on face value I'm not sure about this. Let's go out and talk about people who are knowledgeable about this.'

I went out and started doing some speaking gigs on blockchain. A lot of the people at those gigs would be the 'blockchain junkies'. I put a more sceptical point of view, and say, 'Is there something I'm missing here?' I wasn't going out and saying, 'Hey I know everything and this is all rubbish.' I wanted to explain, 'Well, some people are saying this, but what I see is this. So where is it that I'm wrong? Are these technology problems that we can solve, or are we actually fundamentally trying to confound the laws of physics or ignore how humans operate in society.'

Some things sound good, but rather than just jump on the bandwagon, you must understand the fundamental principles. Exploration, curiosity, and judgement are crucial qualities in the digital age, as is seeking out people who have got a different view. I get out and about a lot. I talk to people. I look at stuff, particularly what I don't understand or I'm not sure about. That is what I go and explore the most so I can come up with a point of view.

WHAT'S HOLDING OUR EXECUTIVES BACK?

One problem that holds executive from taking action is the legacy of technology and the people who manage it. These technology people both come from a different era where they were in control of their overall environment. They had people, they had hardware and they had software. Things were relatively stable and didn't move so fast. They had a lot of money to invest in large scale technology, which tended to be focused on internal processes, not connecting across organisations and customers and 'value-chains', if you want to use a consulting word.

Another barrier is that there's so much other stuff going on every day; we talk about being 'zoomed in' or 'zoomed out'. In a lot of organisations, people are very much zoomed in. If you're the CEO of a public company, you are zoomed in on questions like: 'What are my next quarterly results going to be? What's the share-market doing? What acquisitions are we making?'

They are not asking, 'What opportunities can we create with this changing technology in society and the way that we operate?'

If there has not been any competitive pressure, we often see incumbent organisations not really doing much. Retail in Australia is a good example. Many large Australian retailers didn't fully embrace eCommerce although there has been a lot more action in recent years. We have seen an influx of international retailers in recent years come to our shores. They had built customer bases here through their online channels. So when Amazon decides to open an operation in Australia, suddenly all the retailers are concerned. The problem is that Amazon are digital masters with more than 20 years of experience and the capacity to innovate at a rapid tempo. Organisations that have not embraced digital are now up against the best in class. Not all of the international entrants will succeed but we are seeing an increase in competitive intensity and existing organisations now have to respond.

I say to organisations, 'If you want to harvest wood today, you needed to plant the tree 10 years ago.'

I was talking to a fleet manager, and he asked: 'When will driverless cars become mainstream?' I said to the guy, 'Actually that's the wrong question. You're saying, 'How long do I have to wait to do anything?' My answer would be, the time to do something is today. How do I see this potentially playing out? What role do I want to play, and what can I start doing today that positions me to this in the long-term?'

Look at Deloitte Digital. When I took over running Eclipse (Deloitte Digital's predecessor) I had a lot of persuading to do. I said, 'We are going to do five things every year that have never been done before.' After, just repeating that for five years, people started to believe that, and we started to win these global awards, and people started to say, 'Yeah, it was that narrative, it was that determination, it was putting out the impossible goal, but with definite steps about how we go about achieving it and what do and do not do.'

That's what I'm seeing with organisations; a lot of them wait until they're in the midst of being disrupted before they start doing anything. A good example of this would be broadcast TV. Our commercial networks didn't get serious about streaming until immediately before Netflix arrived in Australia. Interestingly enough it was the public broadcaster, the ABC, which developed its streaming capabilities through the launch of the app iView in 2010. It may have been that the commercial broadcasters couldn't make a business case because it was unclear how the market might play out. A key part of navigating the future is understanding long-term trends and taking action now. It is not a matter of betting the farm; it is about learning and building capability to be prepared for the future.

Big companies are often addicted to the revenue of today. Michael Tushman, **chair of the** program for leadership development at Harvard Business School, came up with the catchphrase: exploit and explore. You need to exploit the opportunities presented by the business of today, however, you also need to explore what the opportunities may be for the business of tomorrow. As he says this is not an either/or choice, it is both/and.

I am an avid reader. I have been influenced by a number of books I read back in the 1990s that still hold true today. I am particularly interested in those books that look at where things might go and cover what they see as the key guiding principles. Three books that really influenced me were:

Being Digital by Nicholas Negroponte (1995) was a window into how the society could operate in a digital future.

New Rules for the New Economy (1997) by Kevin Kelly, founding editor of Wired magazine, outlined what the economic rules might be in a hyper-networked world.

Net Gain by John Hagel (1997) focused on expanding markets through virtual communities and outlined how power was shifting towards customers and the need to build communities.

These three classics are still worth a read today.

When I start thinking about concepts I read in a book, I quickly move from the theory to looking at how to apply them. I'm more the thinker-doer type rather than the thinker or the doer. Some people do, some people think; I sit in the middle of that. I like to test the concepts that I come across to test how they play out in real-world scenarios. We're in a period of history of fundamental change. We've never been as connected as a species or a society, and it's throwing up all sorts of challenges that we've never really had to contemplate before. We're in a hyper-connected world now, highly networked, and we can do things completely differently. Whether it's business, whether it's society, whether it's government, things are playing out in weird and wonderful ways. Often people don't quite understand how all this stuff is happening, but it's happening. Try explaining cryptocurrency to your grandparents and you will get what I mean. I often say if people don't really understand what is happening, you are just as qualified as anybody else to try new things and participate. An old mentor once said to me: if no one knows what to do don't be afraid to get in harm's way and try to work it out.

WHY MANY COMPANIES ARE NOT SUCCEEDING

Most efforts at change take too long, cost too much, and don't recognise the way that values have changed. Most don't recognise the level of experience that people (as consumers) have become used to because, no matter who you are and what your business is, you are no longer competing against the experiences provided by your competitors; your customers expect the sort of experiences provided by Amazon, Apple and Spotify. A fellow futurist Anders Sorman-Nilsson says, 'Today's luxury is tomorrow's expectation'. We need to be able to provide great experiences for customers but also employees. I believe that you can't provide a great customer experience if you don't have great employee experiences. I don't see that many organisations committed to exceptional experiences for customers and employees.

Doing what we used to do doesn't work in this world, and events are not going backwards. They're going to keep moving forward. You've got to increase the tempo of doing and learning. Hone your capacity to try things and learn from them. I meet CEOs who feel like they're swimming in quicksand, and saying, 'We know there's another way. There must be another way. We suspect you know it, or you can help us learn it.'

DO SOMETHING

You've got to learn by doing. You're not going to just analyse this through spreadsheets and aggregating guesses, and doing PowerPoints. That just doesn't move you any closer to your end goal. We've got to learn, we've got to cut our teeth on this, and we are going to make mistakes.

One idea I talk about is 'working out loud'. Let's focus on progress, but let's make sure that everything we do is posted on an internal social network site where we all see what's going on. Anybody who comes in can see where we're at. We can keep moving forward. It's as much a learning exercise from doing as well as getting towards the outcome that you want. And you might be genuinely surprised at what they can achieve.

AIM, FIRE, ADJUST

Aim: Choose a hill to take.

Go back to the outcome: do you want to achieve? Then ask, How are we going to go about this? Is it changing the way that we work? Changing the way we collaborate? Being more transparent? Partnering and working through broader ecosystems rather than trying to do it all ourselves? These are the things that start to come out.

To be a Future Maker, you need to understand not just what's here now, but its trajectory. Where is it heading? The future stuff happens on the edges, on the fringes, not in the centre of your IT department that's delivering the solutions of today. How do we understand what's happening on the edges that we expect to become mainstream? Then ask, 'What does that mean for me? How do I want the future to look? What can I do today to achieve a long-term outcome that I would prefer?'

Once you get to the point that you are ready to try something: Do it, reflect, rethink, go again.

EXPLORE, BE CURIOUS, AND USE YOUR JUDGEMENT

Shareholders are not forgiving. Nor is the media. I've just written a whole chapter telling you that you have to move quickly and be ready to make mistakes. No doubt you are thinking, 'What planet does this guy live on?' But that is why I wrote this chapter. To succeed in our digital future, you must be drawn by curiosity to discover what is over the horizon. Your crew will try to convince you it's the edge of the world, an abyss. You have to lead your company towards it, and keep up the courage of your team. Keep experimenting, spend more time thinking and doing than detailed planning, move fast and work and learn out loud.

TALK TO ME

If you are a doer I'd love to hear you. This chapter is for you. Email me with 'Future Makers' in the subject line and quote a couple of lines from this chapter and I'll make time to listen to what you want to achieve. pewilliams@deloitte.com.au

Peter Williams is a recognised thought leader in innovation and a top digital influencer. He is an Adjunct Professor at RMIT University.

Peter started working with internet technologies in 1993. In 1996 founded an eBusiness Consulting group within Deloitte Australia. He became a partner of Deloitte in 1999 and remained a partner for 14 years. Deloitte is one of the four largest global accounting firms.

Peter became the chief executive of the Eclipse Group, a Deloitte subsidiary, and then founded Deloitte Digital. His current role the 'Chief Edge Officer' at the Deloitte Centre for the Edge. The centre helps businesses profit from emerging technology opportunities.

He is a sought-after speaker, technology provocateur and consultant. He runs a private consultancy firm, Rexster Consulting and is the co-founder of a family history site – Lifetimes.co.

Printed by Libri Plureos GmbH in Hamburg, Germany